Todd Frazier



# Using *MultiSIM*® 6.1: Troubleshooting DC/AC Circuits

## Online Services

**Delmar Online**
To access a wide variety of Delmar products and services on the World Wide Web, point your browser to:

**http://www.delmar.com/delmar.html**
or email: info@delmar.com

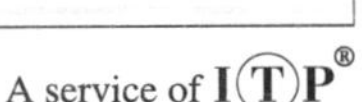

**thomson.com**
To access International Thomson Publishing's home site for information on more than 34 publishers and 20,000 products, point your browser to:

**http://www.thomson.com**
or email: findit@kiosk.thomson.com

A service of I(T)P®

# Using *MultiSIM®* 6.1: Troubleshooting DC/AC Circuits

John Reeder

Merced College
Merced, CA

**Delmar**
Thomson Learning™

Africa • Australia • Canada • Denmark • Japan • Mexico
New Zealand • Philippines • Puerto Rico • Singapore
Spain • United Kingdom • United States

## NOTICE TO THE READER

Publisher does not warrant or guarantee any of the products described herein or perform any independent analysis in connection with any of the product information contained herein. Publisher does not assume, and expressly disclaims, any obligation to obtain and include information other than that provided to it by the manufacturer.

The reader is expressly warned to consider and adopt all safety precautions that might be indicated by the activities herein and to avoid all potential hazards. By following the instructions contained herein, the reader willingly assumes all risks in connection with such instructions.

The Publisher makes no representation or warranties of any kind, including but not limited to, the warranties of fitness for particular purpose or merchantability, nor are any such representations implied with respect to the material set forth herein, and the publisher takes no responsibility with respect to such material. The publisher shall not be liable for any special, consequential, or exemplary damages resulting, in whole or part, from the readers' use of, or reliance upon, this material.

---


**Delmar Staff:**
Business Unit Director: Alar Elken
Executive Editor: Sandy Clark
Acquisitions Editor: Gregory L. Clayton
Developmental Editor: Michelle Ruelos Cannistraci
Editorial Assistant: Jennifer Thompson
Executive Marketing Manager: Maura Theriault
Channel Manager: Mona Caron
Marketing Coordinator: Paula Collins
Executive Production Manager: Mary Ellen Black
Production Manager: Larry Main
Senior Project Editor: Christopher Chien
Art Director: David Arsenault
Technology Project Manager: Tom Smith





Printed in Canada

7 8 9 10 XXX 05 04 03 02

For more information, contact Delmar, 3 Columbia Circle, PO Box 15015, Albany, NY 12212-0515; or find us on the World Wide Web at http://www.delmar.com

**Asia**
Thomson Learning
60 Albert Street, #15-01
Albert Complex
Singapore 189969

**Australia/New Zealand**
Nelson/Thomson Learning
102 Dodds Street
South Melbourne, Victoria 3205
Australia

**Canada**
Nelson/Thomson Learning
1120 Birchmount Road
Scarborough, Ontario
Canada M1K 5G4

**International Headquarters**
Thomson Learning
International Division
290 Harbor Drive, 2nd Floor
Stamford, CT 06902-7477
USA

**Japan**
Thomson Learning
Palaceside Building 5F
1-1-1 Hitotsubashi, Chiyoda-ku
Tokyo 100 0003 Japan

**Latin America**
Thomson Learning
Seneca, 53
Colonia Polanco
11560 Mexico D.F. Mexico

**Spain**
Thomson Learning
Calle Magallanes, 25
28015-Madrid
Espana

**UK/Europe/Middle East**
Thomson Learning
Berkshire House
168-173 High Holborn
London
WC1V 7AA United Kingdom

Thomas Nelson & Sons LTD.
Nelson House
Mayfield Road
Walton-on-Thames
KT 12 5PL United Kingdom





ISBN 07668-1133-6

# Contents


**Preface** . . . . . . . . . . . . . . . . . . . . . . . . . . . . . . **xi**

**CHAPTER 1 Introduction to Electronics Workbench MultiSim®: The Electronics Lab in the Computer** . . . . . . . . . . . . . . . . **1**

Working within the Microsoft Windows™ Environment . . . . . . . . 1

Opening and Using *Electronics Workbench MultiSIM®* . . . . . . . . 1

**CHAPTER 2 Introduction to Electricity and Electronics: Electrical Quantities and Components** . . . . . . . . . . . . . . . . . . . . . **6**

Activity 2.1: Electronic Components . . . . . . . . . . . . . . . . 7

Activity 2.2: The Basic Circuit . . . . . . . . . . . . . . . . . . . 11

Activity 2.3: Changing and Measuring Resistance Values . . . . . 12

Activity 2.4: Changing Voltage Values and Related Circuit Changes . . . . . . . . . . . . . . . . . 14

Activity 2.5: Resistor Color Codes (Three- and Four-Band Resistors) . . . . . . . . . 14

Activity 2.6: Resistor Color Codes (Five-Band) . . . . . . . . . . 15

Activity 2.7: Adjusting and Measuring Potentiometer Resistance in EWB . . . . . . . . . . . . . . . . 16

Activity 2.8: Validating Ohm's Law . . . . . . . . . . . . . . . . 18

Activity 2.9: Calculating Power Consumption . . . . . . . . . . . 20

**CHAPTER 3 Electric Circuits** . . . . . . . . . . . . . . . . . . . . . . . . . **21**

Introduction . . . . . . . . . . . . . . . . . . . . . . . . . . . . . 21

Activity 3.1: Recognizing a Complete Circuit . . . . . . . . . . . 22

Activity 3.2: Recognizing a Series Circuit . . . . . . . . . . . . . 23

Activity 3.3: Recognizing a Parallel Circuit . . . . . . . . . . . . 23

Activity 3.4: Recognizing a Series-Parallel Circuit . . . . . . . . 23

Activity 3.5: Determining Current Paths in a Complex DC Circuit . . . . . . . . . . . . . . . . 24

Activity 3.6: Measuring Voltage in a Circuit . . . . . . . . . . . . 25

**CHAPTER 4 Analyzing and Troubleshooting Series Circuits** . . . . . . . . . . . **28**

Introduction . . . . . . . . . . . . . . . . . . . . . . . . . . . . . 28

Activity 4.1: Recognizing a Series Circuit . . . . . . . . . . . . . 29

Activity 4.2: Measuring Voltage and Current in a Series Circuit . . . . . . . . . . . . . . . . . . . 29
Activity 4.3: Using Kirchhoff's Voltage Law in a Series Circuit . . . . . . . . . . . . . . . . . . . 31
Activity 4.4: Determining Total Resistance in a Series Circuit . . 32
Activity 4.5: Determining Unknown Parameters in Series Circuits . . . . . . . . . . . . . . . . . . . 32
Activity 4.6: Grounding the Circuit . . . . . . . . . . . . . . . . 34
Activity 4.7: Series-Aiding and Series-Opposing Power Sources . . . . . . . . . . . . . . . . . . . 35
Activity 4.8: Unloaded Voltage Dividers and Resistor Power Requirements . . . . . . . . . . . . . . . 36

**CHAPTER 5 Analyzing and Troubleshooting Parallel Circuits . . . . . . . . . . 37**
Introduction . . . . . . . . . . . . . . . . . . . . . . . . . 37
Activity 5.1: Recognizing a Parallel Circuit . . . . . . . . . . . . 38
Activity 5.2: Measuring Voltage and Current in a Parallel Circuit . . . . . . . . . . . . . . . . . . . 39
Activity 5.3: Using Kirchhoff's Current Law in a Parallel Circuit . . . . . . . . . . . . . . . . . . . 40
Activity 5.4: Determining Total Resistance in a Parallel Circuit . . . . . . . . . . . . . . . . . . . 41
Activity 5.5: Determining Unknown Parameters in a Parallel Circuit . . . . . . . . . . . . . . . . . . . 44

**CHAPTER 6 Analyzing and Troubleshooting Series-Parallel Circuits . . . . . . 46**
Introduction . . . . . . . . . . . . . . . . . . . . . . . . . 46
Activity 6.1: Recognizing a Series-Parallel Circuit . . . . . . . . 47
Activity 6.2: Simplifying Series-Parallel Circuits . . . . . . . . . 48
Activity 6.3: Measuring Voltage and Current in a Series-Parallel Circuit . . . . . . . . . . . . . . . 48
Activity 6.4: Loaded and Unloaded Voltage Divider Circuits . . . 50
Activity 6.5: The Effects of Voltmeter Loading . . . . . . . . . . 51
Activity 6.6: Dealing with Bridge Circuits . . . . . . . . . . . . . 52

**CHAPTER 7 Complex Circuits and Network Analysis . . . . . . . . . . . . . . 54**
Introduction . . . . . . . . . . . . . . . . . . . . . . . . . 54
Activity 7.1: Recognizing a Complex Circuit . . . . . . . . . . . 55
Activity 7.2: Mesh Analysis . . . . . . . . . . . . . . . . . . . 55
Activity 7.3: Nodal Analysis . . . . . . . . . . . . . . . . . . . 57
Activity 7.4: The Superposition Theorem . . . . . . . . . . . . . 57
Activity 7.5: Using Thevenin's Theorem. . . . . . . . . . . . . . 58
Activity 7.6: Using Norton's Theorem . . . . . . . . . . . . . . . 59
Activity 7.7: Tee-to-Pi and Pi-to-Tee Conversions. . . . . . . . . 61

**CHAPTER 8 Electrical Power Sources** . . . . . . . . . . . . . . . . . . . . . . . **64**

Introduction . . . . . . . . . . . . . . . . . . . . . . . . . . . 64

Activity 8.1: Using DC Power Sources With Internal Resistance. . . . . . . . . . . . . . . . . . 65

Activity 8.2: DC Power Sources in Series . . . . . . . . . . . . . 67

Activity 8.3: DC Power Sources in Parallel . . . . . . . . . . . . 68

Activity 8.4: Additional DC Power Sources in EWB . . . . . . . 69

Activity 8.5: Introducing the Alternating Current (AC) Power Source . . . . . . . . . . . . . . . . . . 69

**CHAPTER 9 Measuring Direct Current Parameters** . . . . . . . . . . . . . . . **70**

Introduction . . . . . . . . . . . . . . . . . . . . . . . . . . . 70

Activity 9.1: Using the Digital Multimeter (DMM) in DC Circuits . . . . . . . . . . . . . . . . . . . . 71

Activity 9.2: Using the Digital Panel Meters in DC Circuits. . . . 74

Activity 9.3: Internal Resistance of Current and Voltage Meters. . . . . . . . . . . . . . . . . . . 75

Activity 9.4: Calculating the Value of Shunt Resistors for Ammeters . . . . . . . . . . . . . . . . . . 76

Activity 9.5: Calculating the Value of Multiplier Resistors for Voltmeters . . . . . . . . . . . . . . . . . . 77

Activity 9.6: The Oscilloscope as a DC Test Instrument. . . . . . 78

**CHAPTER 10 Electromagnetic Devices** . . . . . . . . . . . . . . . . . . . . . . . **81**

Introduction . . . . . . . . . . . . . . . . . . . . . . . . . . . 81

Activity 10.1: Using the Inductors Found in EWB . . . . . . . . . 82

Activity 10.2: Using EWB Transformers . . . . . . . . . . . . . . 83

Activity 10.3: Using the EWB Relay Component. . . . . . . . . . 84

Activity 10.4: Using the EWB DC Motor . . . . . . . . . . . . . . 85

**CHAPTER 11 Alternating Voltage and Current**. . . . . . . . . . . . . . . . . . . . **86**

Introduction . . . . . . . . . . . . . . . . . . . . . . . . . . . 86

Activity 11.1: AC Waveforms . . . . . . . . . . . . . . . . . . . . 87

Activity 11.2: Measuring Time and Frequency of AC Waveforms . . . . . . . . . . . . . . . . . 87

Activity 11.3: Measuring RMS, Average, Peak, Peak-to-Peak, and Instantaneous Values of AC Waveforms . . . 90

Activity 11.4: Phase Relationships between Waveforms . . . . . . 91

Activity 11.5: Harmonic Relationships between AC Signals . . . . 92

Activity 11.6: Characteristics of EWB Measuring Instruments in AC Circuits . . . . . . . . . . . . . . . . . . . 92

Activity 11.7: Voltage, Current, Resistance, and Power in Resistive AC Circuits . . . . . . . . . . . . . . . 94

**CHAPTER 12 Inductance and Inductive Reactance** . . . . . . . . . . . . . . . . . **96**
Introduction . . . . . . . . . . . . . . . . . . . . . . . . . . . . 96
Activity 12.1: Measuring Inductance with an Inductance Test Circuit . . . . . . . . . . . . . . . . . . . . 97
Activity 12.2: Calculate and Measure Inductance . . . . . . . . . . 98
Activity 12.3: Calculating and Measuring Inductive Reactance in an AC Circuit . . . . . . . . . . . . . . . . . . 99
Activity 12.4: Phase Relationships in Inductive Circuits . . . . . . 101
Activity 12.5: The Q of an Inductor and Power Losses in Coils . . 102

**CHAPTER 13 Resistive-Inductive Circuits** . . . . . . . . . . . . . . . . . . . . . . . **104**
Introduction . . . . . . . . . . . . . . . . . . . . . . . . . . . . 104
Activity 13.1: Series Resistive-Inductive (RL) Circuits . . . . . . . 105
Activity 13.2: Parallel Resistive-Inductive (RL) Circuits . . . . . . 108
Activity 13.3: Series-Parallel Resistive-Inductive (RL) Circuits . . 111
Activity 13.4: Pulse Response and L/R Time Constants of RL Circuits. . . . . . . . . . . . . . . . . . . . . 113

**CHAPTER 14 Capacitance and Capacitive Reactance** . . . . . . . . . . . . . . **116**
Introduction . . . . . . . . . . . . . . . . . . . . . . . . . . . . 116
Activity 14.1: Testing Capacitors with a Capacitor Test Circuit . . 117
Activity 14.2: Calculating and Measuring Total Capacitance . . . . 119
Activity 14.3: Calculating and Measuring Capacitive Reactance in an AC Circuit . . . . . . . . . . . . . . . . . . 120
Activity 14.4: Phase Relationships in Capacitive Circuits. . . . . . 122

**CHAPTER 15 Resistive-Capacitive Circuits** . . . . . . . . . . . . . . . . . . . . . . **125**
Introduction . . . . . . . . . . . . . . . . . . . . . . . . . . . . 125
Activity 15.1: Series Resistive-Capacitive (RC) Circuits . . . . . . 125
Activity 15.2: Parallel Resistive-Capacitive (RC) Circuits . . . . . 129
Activity 15.3: Series-Parallel Resistive-Capacitive (RC) Circuits. . 131
Activity 15.4: Pulse Response and Time Constants of RC Circuits . . . . . . . . . . . . . . . . . . . . 133

**CHAPTER 16 Resistive-Inductive-Capacitive Circuits** . . . . . . . . . . . . . . . **136**
Introduction . . . . . . . . . . . . . . . . . . . . . . . . . . . . 136
Activity 16.1: Series Resistive-Inductive-Capacitive (RLC) Circuits . . . . . . . . . . . . . . . . . . . . 137
Activity 16.2: Parallel Resistive-Inductive-Capacitive (RLC) Circuits . . . . . . . . . . . . . . . . . . . . 141
Activity 16.3: Series-Parallel Resistive-Inductive-Capacitive (RLC) Circuits . . . . . . . . . . . . . . . . . . . . 143
Activity 16.4: Series and Parallel Resonant Circuits . . . . . . . . 145

**CHAPTER 17 Transformer Circuits . . . . . . . . . . . . . . . . . . . . . . . . . . 148**
Introduction . . . . . . . . . . . . . . . . . . . . . . . . . . . . . 148
Activity 17.1: Transformer Step-Up and Step-Down Circuits. . . . 148
Activity 17.2: Transformer Buck/Boost Circuits . . . . . . . . . . 150

**CHAPTER 18 Passive Filter Circuits . . . . . . . . . . . . . . . . . . . . . . . . . . 152**
Introduction . . . . . . . . . . . . . . . . . . . . . . . . . . . . . 152
Activity 18.1: The EWB Bode Plotter . . . . . . . . . . . . . . . . 153
Activity 18.2: Low-Pass Filter Circuits . . . . . . . . . . . . . . . 155
Activity 18.3: High-Pass Filter Circuits . . . . . . . . . . . . . . . 157
Activity 18.4: Band-Pass Filter Circuits . . . . . . . . . . . . . . . 158
Activity 18.5: Band-Reject, Bandstop, and Notch Filter Circuits . . 159

# Preface

## Approach

This workbook is designed to teach the student how to virtually measure and troubleshoot electronic circuits created within the *Electronics Workbench MultiSIM®* Version 6.11 software environment and to reinforce DC/AC theory learned in the classroom. The computer and the computer monitor become the electronics workbench. Students using this manual must have *MultiSIM* Version 6.11 installed on their computers to be able to operate the provided software projects. These software projects are stored on the CD that accompanies this text.

An advantage to the virtual laboratory approach to electronics is the low cost of the software package in comparison to the expenses required to establish an electronics laboratory with all of the necessary test equipment and the related facility costs. Virtual electronic circuits can be modified easily on the monitor screen and circuit analysis is also easily obtained as circuits are modified.

Circuit troubleshooting is an integral component of the software package. It is relatively easy for the instructor or textbook author to install faults such as shorts, leakage, and opens into the circuit for the electronics student to locate. The troubleshooting exercises will provide the student with the confidence and skills necessary to troubleshoot circuits constructed on the laboratory workbench.

## System Requirements

Pentium 166 or greater PC
Windows 95/98/NT
32 MB RAM (64 MB RAM recommended)
100–250 MB hard disk space (min.)
CD-ROM drive
800 x 600 minimum screen resolution

## Organization

**Activity** sections are located in each chapter that break the overall subject of the chapter down into smaller blocks. The individual software projects are related to subtopics within the larger topic.

Within each activity, **circuit files** progressively provide individual projects related to the subject material of the chapter. Most subjects are touched upon as progress is made through the projects. A final circuit file for many of the activity sections is a **troubleshooting** problem/s.

## Circuit Files

The CD that accompanies the text contains all of the circuit files in this book. They are pre-built and ready to be used with *Electronics Workbench MultiSIM®*. There are over 250 circuits available.

Circuit files follow the DOS system of nomenclature. There is a maximum of eight digits in each circuit name, but unlike previous versions of *Electronics Workbench®*, the circuit name is not followed by a DOS extension. The first two numbers of the circuit file name represent the chapter in the book followed by a dash. After the dash, the next two numbers represent the activity within the chapter. The letter, and occasionally a letter followed by a number, represents the sequence of events within an activity.

## Instrumentation

Test instruments are accessed through the **Instruments** button on the Design Bar above the circuit workspace. A left-click on this button will cause the instruments toolbar to appear at the bottom of the screen. You will use the DMM, the oscilloscope, and the Bode plotter with this workbook. The DMM and the oscilloscope are common instruments found on the test bench. The Bode plotter is a virtual instrument found only in a software program, it is not a real world instrument. The Bode graphic plot, a diagram of voltage amplitude of a circuit in reference to the circuit frequency, is commonly used in electronics to better understand circuit operation as frequency changes. The Bode plotter used in EWB displays the Bode plot of a circuit as if it were a real test instrument and it is similar to a spectrum analyzer, an advanced piece of test equipment found in some electronics test labs.

## Additional Resources

Delmar Thomson Learning provides support for electronics instructors on their web site at www.electronictech.com. Answers to questions and problems in the text will be provided in a password-protected location accessible only to instructors.

Electronics Workbench *MultiSIM*® can be purchased at your college bookstore or go to www.electronictech.com for information.

## About the Author

John Reeder, A.A., B.A., M.S., is the department head of Electronics Technology at Merced College and also taught high school electronics for nine years. Prior to becoming a teacher, he worked for 28 years in the electronics and electrical industries as technician, electrician, and engineer. He has been using *Electronics Workbench*® for many years as a supplement to help students gain a better understanding of their electronics material, written projects, and created files using *Electronics Workbench*®. He recently completed work as a beta-tester for Interactive Image Technologies in reviewing *MultiSIM.* At Merced College, he has developed course curriculum using EWB as the software core of the overall electronics program.

## Acknowledgments

I would like to express my grateful appreciation to the publishing team at Delmar for patiently working with me and helping me over the hurdles of a first book. The members of that team consist of Michelle Ruelos Cannistraci, Greg Clayton, and Jennifer Thompson. Also, I would like to thank Larry Main and David Arsenault at Delmar who helped me with software and graphics problems.

I would also like to express my grateful appreciation to the software team at Interactive Devices, Scott Duncan, Roman Bysh, Luis Alves, and Leo who helped me through the difficult moments when the software wasn't responding to my efforts. All worked out well and the result is this finished product.

I want to thank my partner in electronics instruction at Merced College, Bill Walls, who reviewed the initial draft of the book. His review of the initial draft is greatly appreciated.

I also want to thank my students who provided the original impetus to get me started in writing and developing EWB files for their use. Their enthusiasm over the book and the EWB projects is wonderful.

Last, but not least, I have to acknowledge the most important member of my team here in Merced, my wife Barbara, who put up with the long hours spent on the computer over a summer break, winter break, weekends, and every other spare moment. She always provided the right touch at the right moment and kept me going at critical moments.

# 1. Introduction to Electronics Workbench®: The Electronics Lab in the Computer

> ***References***
> *Electronics Workbench®, MultiSIM* Version 6
> *Electronics Workbench®, MultiSIM* Version 6 User's Guide

**Objectives** After completing this chapter, you should be able to:

- Open and use the *Electronics Workbench® MultiSIM* program.
- Open EWB component and instrument libraries.
- Construct basic electronics circuits in EWB.

## Working within the Microsoft Windows™ Environment

The *Electronics Workbench® MultiSIM* "electronics laboratory simulation" software currently uses the Microsoft Windows™ 3.xx, 95 or 98 operating systems that are installed on many desktop computers found in homes, schools, and industry. On the workspace of the computer screen there should be a *MultiSIM* icon that will activate the installed *Electronics Workbench®* software program. When this icon is activated, the EWB program will load and wait for your command: a virtual electronics laboratory.

## Opening and Using *Electronics Workbench MultiSIM®*

### Introducing *MultiSIM* Version 6

When the program is first activated, the display presents a blank circuit window entitled "Circuit 1." This window can be used to construct a circuit of choice, to display a "pre-built" circuit found within the program file system, or to display a circuit file found in another location such as a CD or floppy disk. When looking for circuits within the file system or another location, start by clicking on **File** and **Open** with the mouse as with any Windows™

program. Then open the file folder where the circuit file is located. This location can be in the program folder on the computer hard drive or in an extraneous source such as a CD or a floppy disk. The circuit files that accompany this text are located on the accompanying CD and can be installed on the computer hard drive. The file system used within EWB is the same as the Microsoft Windows™ file system and can be manipulated using the same methods and techniques as File Manager or Explorer. If the project files are installed on the computer hard drive, they should be located in the MultiSIM folder as shown in Figure 1-1 (DC/AC Circuits).

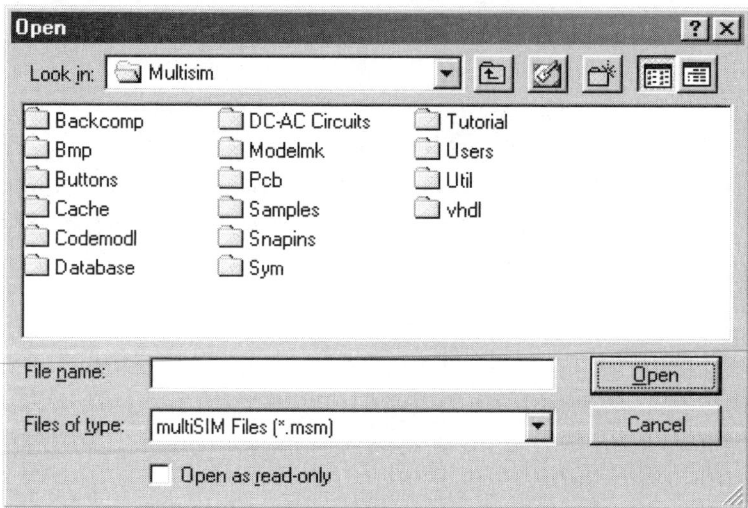


**Figure 1-1** The Basic Menu of EWB 5

*Electronics Workbench MultiSIM®* is best defined as an electronics workbench in a virtual laboratory. Practically every circuit studied in lower level electronics courses can be constructed and tested using this powerful software application on a computer. In a real laboratory environment, EWB can be used to verify and compare concepts concerning electronic circuits constructed on the workbench. This workbook is primarily designed to reinforce subject material covered in any electronics textbook at the DC/AC level used in secondary and post-secondary education.

## Component and Instrument Libraries *(MultiSIM)*

Launch EWB and observe the blank circuit window. The **Component Toolbars** are located to the left of the circuit window. There are fourteen component icons (parts bins) on this toolbar as shown in Figure 1-2. This figure is

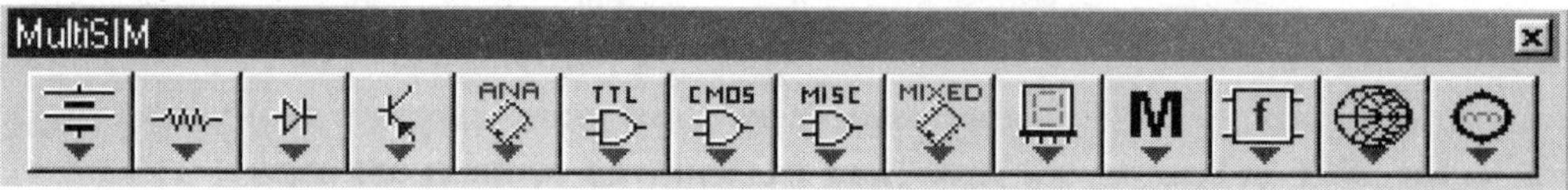


**Figure 1-2** The EWB *MultiSIM* Component Toolbar

shown in the horizontal position rather than its normal vertical location to the left of the screen.

The MultiSIM user can access the individual libraries with the mouse using normal drag-and-drop techniques. Each of these libraries can be identified, regarding their contents, by resting the mouse pointer on a specific library and waiting for the drop-down **Tool Tip** to appear. When going though this text, the primary component libraries that will be used are **Sources**, **Basic**, **Indicators**, and **Miscellaneous** libraries.

Using the mouse, open each of the component libraries and observe their contents, contents that will become very familiar while progressing through the study of DC/AC electronics technology with the help of EWB. Locate the position of the **Sources** (Figure 1-3) and **Basic** (Figure 1-4) libraries, the **Indicators** (Figure 1-5) library, and finally the **Miscellaneous** (Figure 1-6) library. Use the drop-down **Tool Tip** to verify the library locations.

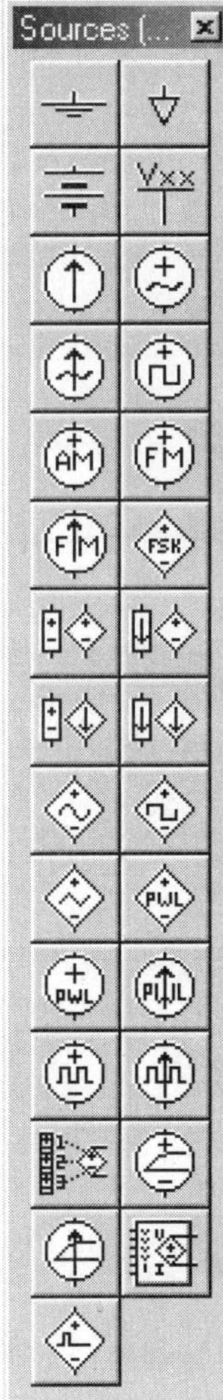


**Figure 1-3** The Sources Library

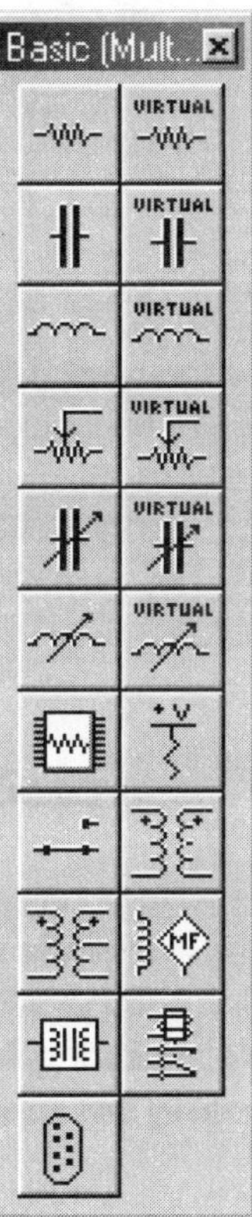


**Figure 1-4** The Basic Library

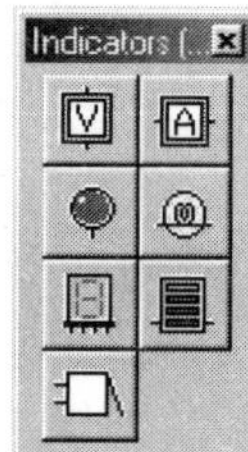

**Figure 1-5** The Indicators Library

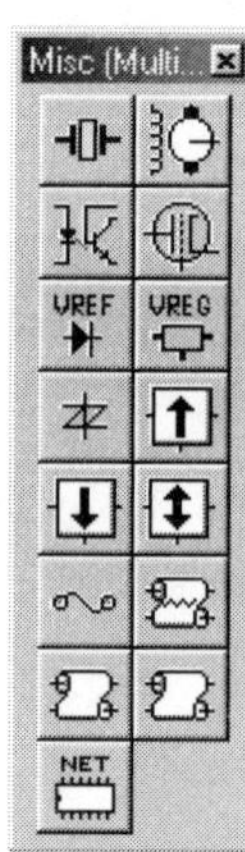


**Figure 1-6** The Miscellaneous Library

EWB *MultiSIM*® contains typical test equipment that is found in an electronics laboratory (virtual test equipment). This inventory of virtual test equipment includes a multimeter, a function generator, an oscilloscope, and a Bode plotter that will be used in this text. There is also other test equipment that will not be used at this point in the study of electronics. Also, within the **Indicators** library there are digital voltmeters, digital ammeters and a red voltage probe, each of which may be used as many times as needed in a circuit. The instruments within the **Instruments** library can be used in more than one location in a circuit. For example, there can be more than one multimeter on the circuit workspace.

## Using the Help function in EWB *MultiSIM*®

Any time you need to know more about a component or an instrument, the first step is to drag the component or instrument out onto the workspace. Then right-click on the component and choose the help function from the bottom of the menu. You can also obtain information in the User Guide that comes with the software.

## Saving your work

When you have completed your project or the work period is nearing an end, you can save your work on a floppy disk or on the hard drive of the computer (with your instructor's permission). To save your file, open the drop down **File** menu at the upper right of the taskbar. Left-click on **Save As...** and the menu shown in Figure 1-7 will appear. You can save the file in this location under "your name.ewb" or on a floppy disk (preferred). To save to a floppy disk, left-click on the arrow to the right of the MultiSIM window (see Figure 1-8). Left-click on the 3-1/2″ Floppy (A:) and save to the disk.

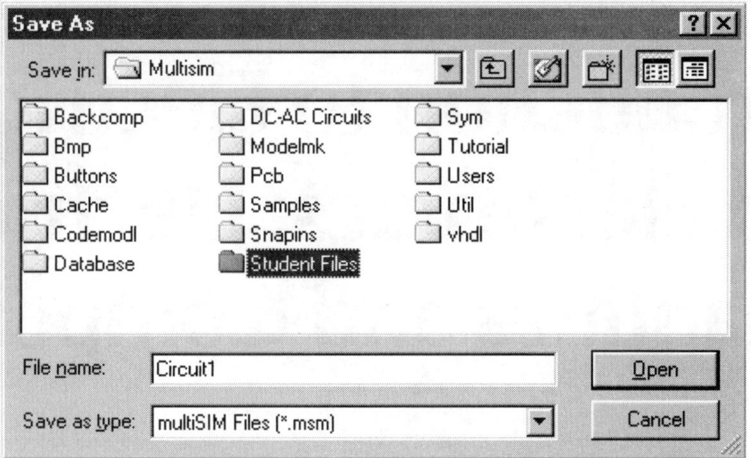


**Figure 1-7** The **Save As** Menu

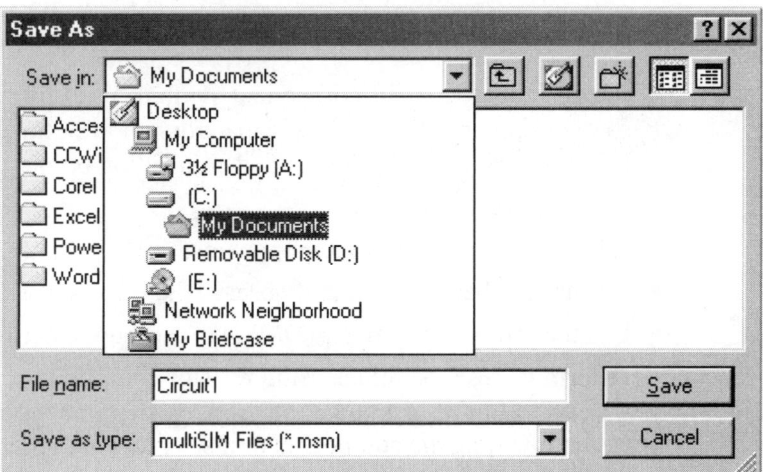


**Figure 1-8** Saving to a Floppy Disk

# 2. Introduction to Electricity and Electronics: Electrical Quantities and Components

***References***
*Electronics Workbench®, MultiSIM* Version 6
*Electronics Workbench®, MultiSIM* Version 6 Study Guide

**Objectives** After completing this chapter, you should be able to:

- Recognize basic EWB component symbols used for resistors, DC power sources (batteries), inductors, capacitors, diodes, transistors, transformers, switches, fuses, and connection points.
- Activate prebuilt circuits.
- Recognize and use EWB components.
- Construct a basic circuit and observe the effects of DC current flow.
- Use the virtual multimeter (DMM) to measure electrical quantities.
- Determine resistor values using resistor color codes.
- Vary the resistance of potentiometers in EWB.
- Validate and use Ohm's law.
- Calculate power consumption.

## Introduction

The knowledge of electronics begins with a thorough understanding of the foundational phenomena of charge, voltage, current, resistance, capacitance, inductance, and power. All of electronics hangs on these root concepts. Electronics circuits exhibit all of these basic phenomena in varying degrees and are dependent upon the value of the circuit components and the amount of voltage applied to the circuit by the voltage source.

Electronic components are the basic building blocks of a circuit. If you look at the resistor, as an example, you discover that it is a component that exhibits certain electrical characteristics with the amount of resistance being identified through a color code system (in the case of most resistors). This you will study. Other components will be studied and explained as you progress through the text.

The circuit diagram, or schematic diagram, can be defined as a symbolic representation of an electrical/electronic circuit. Through the schematic diagram, the physical configuration of a circuit can be determined and then constructed according to the diagram, subject to physical constraints.

## Activity 2.1: Electronic Components

**Electronic components** are the manufactured parts used to construct electronics circuits. These components range from small parts such as transistors, to very complex electronic assemblies. The electronic assemblies may, within themselves, contain many individual components and integrated circuits that became a part of these more complex assemblies in the manufacturing process. Various small components and even very complex manufactured circuits and assemblies become the basic building blocks of ever more complex electronic equipment and systems.

1. **Resistors** are components that are used to oppose or limit current flow in electronic circuits. On the EWB workspace, in close proximity to the resistor symbol, a component designator, a resistance value, and special labels may be displayed. In Figure 2-1 for example, the resistor component symbols have the component designator (R1) and the resistance value (1k) below the components. The symbol on the left is a basic resistor with a menu of values, the symbol in the center is a virtual resistor where the component value can be individually adjusted, and the symbol on the right is a variable resistor, also known as a potentiometer. The potentiometer is widely used as a volume control in audio equipment.

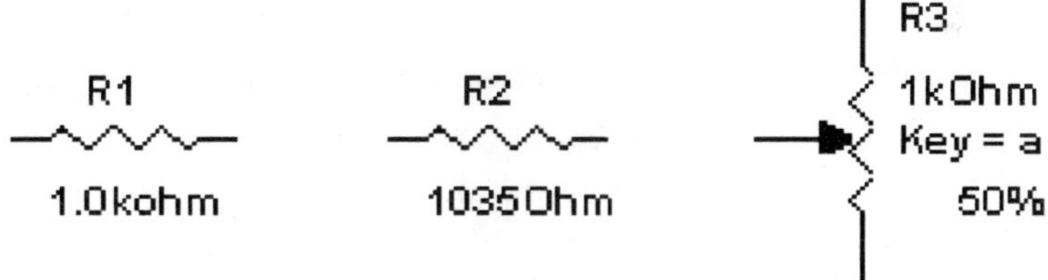


**Figure 2-1** EWB Resistor Symbols

2. Now is the time to begin to use EWB. Open circuit file **02-01a**. It contains two resistors (R1 and R2). A double left-click directly on either of the components will bring up the **Component Properties** menu. Using the **Value** tab, change the values of the resistors from 1 kΩ to 10 kΩ, left-click on **Replace** and notice that the displayed values have been changed from 1 kΩ to 10 kΩ. A single right-click on the component (not on the letters, but right on the component) will bring up a menu with other options such as rotate or delete. This method of changing component settings and values is used throughout the EWB program for most of the components. Choose one of the **Rotate** options and the component will rotate 90 degrees.

3. Now, open the **Basic** menu on the toolbar (the icon with a resistor symbol), left-click on the virtual resistor symbol, then left-click on the workspace, and a resistor will appear. Open the component menu for this new component, select the **Display** tab, and click off the check mark on **use Schematic Options global settings** (notice that **labels, values**, and **reference ID** are checked; these can be turned off also), left-click on **OK** and the new resistor should now have the designation of R3 above it. Use the various menu options, components, and the component menu to help you become more familiar with the operation of the EWB program.

4. **DC Power Sources (Batteries)** provide DC power for the circuits and are represented in EWB by the battery symbol. The DC power sources may also have the component designator, a DC voltage value, and special labels displayed near the component. In Figure 2-2, the component designator (**V1**) and the DC voltage value (**12V**) are shown to the right of the component symbol. There are several other types of power source symbols in EWB that will be introduced later.

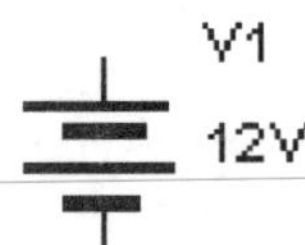


**Figure 2-2** The EWB Battery Symbol

5. Open circuit file **02-01b**. Open the component menu (double left-click) and change the value of the power source to 50 V. Go to the **Display** tab and activate **Show reference ID**, click on **OK**, and the designation **V1** will now be displayed.

6. The **Ground** (common) symbol, displayed in Figure 2-3, is a connection point in most electronic circuits and represents a common point of reference for the circuit. This common point is used for voltage measurements as a point of reference. When constructing an EWB circuit, you have to make sure that you always place a ground somewhere in the circuit (particularly in AC circuits), to ensure that the test instruments and the circuit work properly.

**Figure 2-3** The EWB Ground Symbol

7. **Inductors** consist of coils of wire and store electrical energy in an electromagnetic field. Inductors may also be called coils. In EWB, inductors may have a component designator, an inductance value, and special

labels whenever they are needed. In Figure 2-4, the component designators (L1 and L2) and the inductance value (1 mH) are shown above both of the components. The component symbol to the left is the standard inductor (coil) with a standard menu of values, the center component is a virtual inductor with the component value being set by the user, and the component symbol to the right is a variable inductor.

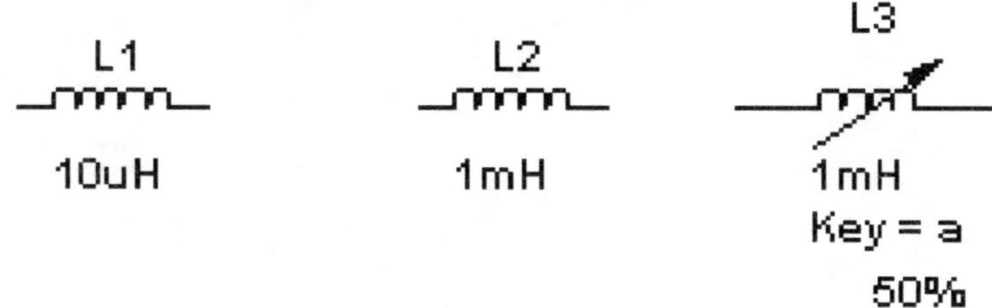


**Figure 2-4** The EWB Inductor Symbols

8. Open circuit file **02-01c**. Open the component menus for each coil and change the value of the inductors to 100 mH. Go to the **Display** tab for each coil, activate **Show reference ID**, click on **OK** and the designation L1 or L2 will be displayed.

9. **Capacitors** consist of two conducting plates separated by an insulator. They store electrical energy in an electrostatic field. In EWB, capacitors may have a component designator, a capacitance value, and special labels when needed. In Figure 2-5, the component designators (C5, C6, and C7) and the capacitance values are shown below the component symbols. The component symbol to the left is a standard component selected from a menu (the + indicates polarity), the center component is a virtual capacitor symbol with the component value being set by the user, and the capacitor symbol to the right is a variable capacitor.

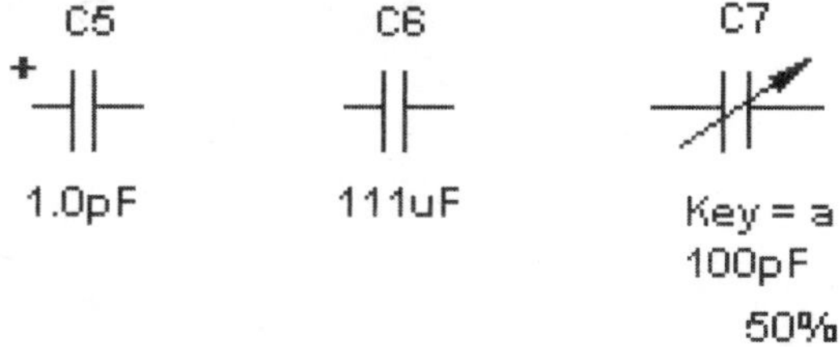


**Figure 2-5** Capacitor Symbols in EWB

10. Open circuit file **02-01d**. Open the component menu and change the values of the capacitors to 100 μF. Go to the **Display** tab for each capacitor and activate **Show reference ID**, click on **OK** and C1 or C2 will be displayed.

11. **Transformers** consist of two or more coils of wires (inductors) whose electromagnetic fields have an interacting relationship due to the proximity of

their windings. In EWB, a transformer may have a component designator and a special label when necessary. In Figure 2-6, the component designators (T1 and T2) are shown near the components.

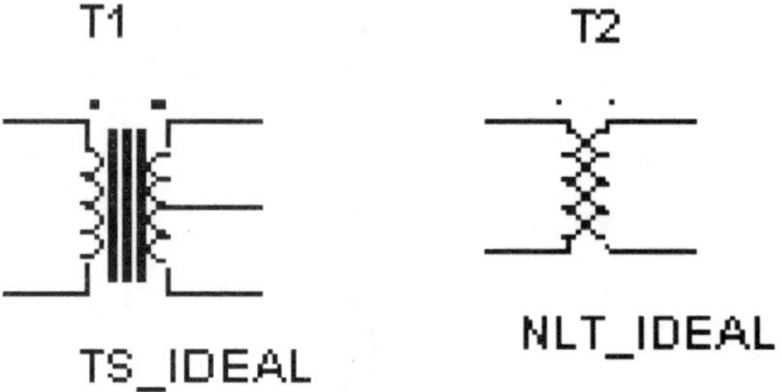


**Figure 2-6** The EWB Transformer Symbols

12. Open circuit file **02-01e**. Double left-click on each of the transformers and view the menu options. Notice that under the **Label** tab, you can change the designation of the transformer from its default setting of $T_1$ or $T_2$. Also, you can assign a label to the component.

13. **Switches** are components that open or close electrical circuits and provide a path for current flow. In EWB, switches have a component designator and special labels when needed. In Figure 2-7, the component designator (S1) is shown above the switch symbol. In EWB a switch is more than a symbol, it will actuate a virtual circuit. That is the purpose of the [Key = Space] designation below the symbol indicating that the **Space bar** on the computer keyboard will virtually activate the switch. It is possible to change the actuating key from [Space] to any other key if desired. If there are a number of switches on the workspace, each one of them can be actuated by a different key on the keyboard or all can be activated by the same key. (There are additional switches located in the Electro-Mechanical library for industrial circuits).

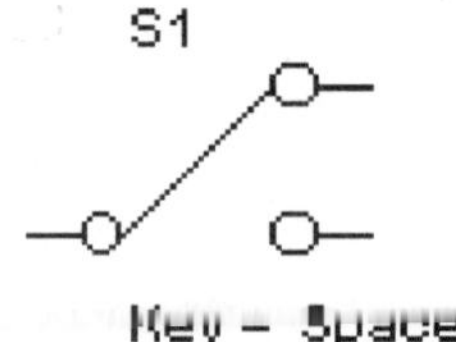


**Figure 2-7** The EWB Switch Symbol

14. Open circuit file **02-01f**. Notice that both switches have the designator [Space]. Toggle the [Space] bar on the keyboard and observe that both switches toggle. Open the **component properties** menu, go to the **Display** tab for each switch and activate **Show reference ID**, click on **OK** and the left switch is designated as "S1" and the right switch is designated as "S2". Now open the menu for Switch 2, open the **Value** tab and change the designation [Space] to A. Return to the workspace and notice that the A key on the keyboard now toggles Switch 2.

15. **Fuses** are used to protect electronic circuits and are located in the **M** (miscellaneous) library near the bottom of the tool bar. In Figure 2-8, the component designator (U1) and the fuse value (1A) are shown above and below the fuse symbol used in EWB. On a "real world" schematic, the designator **F**, for fuse, is commonly used.

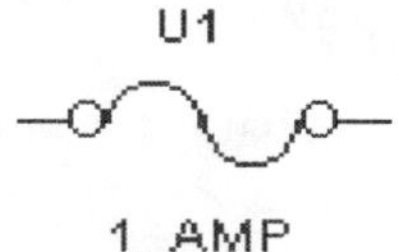


**Figure 2-8** The EWB Fuse Symbol

16. Open circuit file **02-01g**. Notice that the fuse is installed in a circuit, has a value of 5A and is designated as "U1." Activate the circuit. The meter reads about 5A. This circuit is acting normally.

17. Open the fuse menu and change the fuse value to 1A. Reactivate the circuit and observe that the fuse "blows." Return the value of the fuse to 5A, reactivate the circuit again, and the circuit will return to normal operation.

18. There are many other component symbols used in EWB *MultiSIM*®. We will discuss these additional symbols and use them as we progress through the textbook and it becomes necessary to use these additional symbols in more advanced circuits.

## Activity 2.2: The Basic Circuit

1. The minimum requirements for a basic circuit consist of a voltage source, a load (in this case a lamp, which is a resistive device), a control device (such as a switch), and the connecting means (such as wires or a printed circuit board). Figure 2-9 contains all of the necessary components that constitute such a basic circuit.

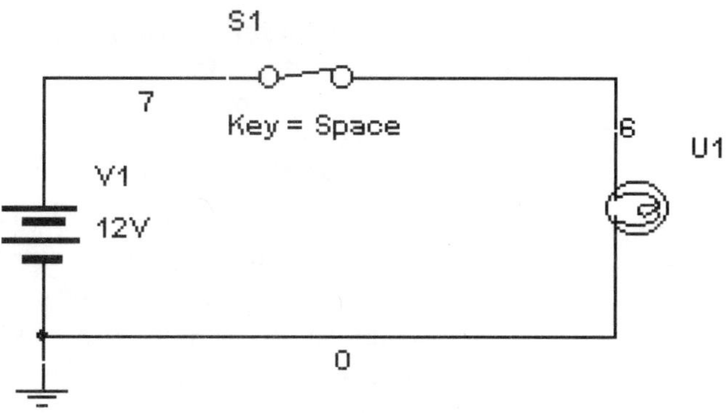


**Figure 2-9** A Basic Circuit

2. Open circuit file **02-02**. In this file a basic circuit constructed on the EWB workspace is displayed.

3. Activate the circuit and close Switch 1 so that current flows in the circuit.

   What is the value of voltage applied ($V_A$) to the circuit? $V_A$ = 12 V. The term $\mathbf{V_A}$ designates voltage applied to a circuit by the power source.

4. What is the amount of current ($I_T$) flowing in the circuit as indicated by meter $M_1$? $I_T$ = .833 A. $\mathbf{I_T}$ designates total current flowing in the circuit.

## Activity 2.3: Changing and Measuring Resistance Values

1. When you change the value of a resistor in a circuit there is a proportional change in the amount of current flowing in the circuit.

2. Open circuit file **02-03a**. Activate the circuit and determine the value of $V_A$. Sometimes the value of $V_A$ is displayed to the right of the voltage source and other times you have to determine $V_A$ by the use of circuit parameters or a test instrument. These are the various means necessary to determine the value of the voltage in the circuit being tested. In this circuit, the voltage is indicated to the right of the voltage source ($V_A$).

   Displayed $V_A$ = 12 V. Determine the value of $I_T$ by double-clicking on the ammeter (XMM1) and reading the circuit current.

   Measured $I_T$ = 12 A.

3. Change the value of $R_1$ to 2 Ω. The use of $\mathbf{R_1}$ designates resistor number 1 in the circuit. Activate the circuit and determine the new value of $I_T$.

   Measured $I_T$ = 6 A.

4. Predict the value of $I_T$ if $R_1$ is changed to 6 Ω. Predicted $I_T$ = 2 A.

5. Change the value of $R_1$ to 6 Ω and verify the prediction. Did they match?

   Yes ________ or No ________.

6. Open circuit file **02-03b**. In this activity the multimeter (DMM) will be used to measure resistance. The two DMM settings to be highlighted are the "Ω" symbol representing resistance and the straight line which represents DC. Move the red probe presently connected to $R_1$ to $R_2$ and then to $R_3$, and so on, then enter the resistance value for each resistor in Table 2-1. The value of resistor $R_1$ is already entered as an example of how to enter the data.

| | Measured Value | | | Measured Value |
|---|---|---|---|---|
| $R_1$ | 15 kΩ | | $R_6$ | 560 kΩ |
| $R_2$ | 220.0kΩ | | $R_7$ | 2.2KΩ |
| $R_3$ | 4700 kΩ | | $R_8$ | 1.2MΩ |
| $R_4$ | 47 Ω | | $R_9$ | 33kΩ |
| $R_5$ | 1kΩ | | $R_{10}$ | 750 Ω |

**Table 2-1** Measuring Resistance

## • *Troubleshooting Problems:*

7. Open circuit file **02-03c**. In this exercise, you will measure the resistance of a group of resistors. In the group, three of the resistors will have a measurement that disagrees with the schematic values. Start by entering the schematic values in Table 2-2 and then measure the resistance of each component and enter that data in the table. Locate the "bad parts" from the obvious discrepancies in resistance values using the data entered into the table.

| | Schematic Value | Measured Value |
|---|---|---|
| $R_1$ | 75kΩ | 75kΩ |
| $R_2$ | 560Ω | 560Ω |
| ✗ $R_3$ | 470Ω | 99.9 |
| $R_4$ | 68Ω | 68Ω |
| ✗ $R_5$ | 1kΩ | 0. Ω |
| $R_6$ | 120Ω | 120 Ω |
| $R_7$ | 1.8kΩ | 1.8kΩ |
| $R_8$ | 1.2MΩ | 1.2kΩ |
| ✗ $R_9$ | 33kΩ | -r- |
| $R_{10}$ | 100 Ω | 100Ω |

**Table 2-2** Locating Defective Resistors

8. Which three components have an incorrect value? R 3, R 5, and R 9 are the three resistors that measure incorrectly according to the schematic values, the rest agree with the schematic and are correct.

## Activity 2.4: Changing Voltage Values and Related Circuit Changes

1. In Activity 2.2, Steps 1 through 5, you learned that by changing resistance in a circuit there was a corresponding change in the value of current flow. The next step is to investigate what happens when the value of voltage applied to a circuit is changed.

2. Open circuit file **02-04**. Activate the circuit. What is the stated value of $V_A$? $V_A$ = 12 V. Double-click on the current meter and determine the value of $I_T$? $I_T$ = 12 A.

3. Change the value of $V_1$ to 10 V. Activate the circuit. What is the new value of $I_T$? Measured $I_T$ = 10 A.

4. Predict the value of $I_T$ if $V_1$ is changed to 5 V. Predicted $I_T$ = 5 A.

5. Now change the value of $V_1$ to 5 V and verify the prediction. Did they match? Measured $I_T$ = 5 A.

## Activity 2.5: Resistor Color Codes (Three- and Four-Band Resistors)

1. Historically, resistors have been found in three-band (20% tolerance) and four-band color code configurations.

2. Open circuit file **02-05**. Activate the circuit and determine the resistance value of the ten resistors.

R1 R6
R2 R7
R3 R8
R4 R9
R5 R10

3. The ten resistors shown are three- and four-band resistors. Determine the color codes for the resistors and enter the answers in Table 2-3. Remember that 20% resistors do not have a fourth color band.

| | Color Band 1 | Color Band 2 | Color Band 3 | Color Band 4 |
|---|---|---|---|---|
| $R_1$ | Brown | Black | Red | Gold |
| $R_2$ | Brown | Black | Orange | |
| $R_3$ | Red | Yellow | Brown | Gold |
| $R_4$ | Orange | Orange | Yellow | Red |
| $R_5$ | Yellow | Violet | Black | Silver |
| $R_6$ | Brn | Blk | Red | Red |
| $R_7$ | Brn | Blk | Grn | |
| $R_8$ | Gray | Red | Brn | Gold |
| $R_9$ | Org | Org | Blk | Silver |
| $R_{10}$ | Violet | Grn | Org | Silver |

**Table 2-3** Three- and Four-Band Resistor Color Codes

## Activity 2.6: Resistor Color Codes (Five-Band)

1. Color-coded resistors are also found in several five-band configurations, one type with the fifth band indicating tolerance (precision resistors) and the second type with the fifth band indicating reliability (specified reliability factor resistors). In this study, five-band **precision** resistors will be studied.

2. Open circuit file **02-06**. The resistors shown are five-band resistors. Determine the proper color codes for the ten resistors and enter the answers in Table 2-4.

| | Color Band 1 | Color Band 2 | Color Band 3 | Color Band 4 | Color Band 5 |
|---|---|---|---|---|---|
| $R_1$ | | | | | |
| $R_2$ | | | | | |
| $R_3$ | | | | | |
| $R_4$ | | | | | |
| $R_5$ | | | | | |
| $R_6$ | | | | | |
| $R_7$ | | | | | |
| $R_8$ | | | | | |
| $R_9$ | | | | | |
| $R_{10}$ | | | | | |

**Table 2-4** Five-Band Resistors

## Activity 2.7: Adjusting and Measuring Potentiometer Resistance in EWB

1. Potentiometers are variable resistors and, when they are used in EWB, can be varied by the keyboard. As previously stated, the setting of a potentiometer (pot) can be changed by depressing the key on the keyboard that is related to that potentiometer. For example, the resistance of $R_1$ in Figure 2-10 can be decreased by depressing the key "**r**" and increased by depressing the key "**R**". $R_2$ uses "**s** and **S**", $R_3$ uses "**t** and **T**", and so on.

2. To determine the resistance at a particular setting of a potentiometer in EWB, multiply the total resistance of potentiometer by the percentage displayed to the right of the potentiometer. For instance, the potentiometers in Figure 2-10 are all set at the 50% point in total wiper travel and their total resistance from the bottom terminal to the wiper would be one half (50%) of their total resistance. The percentage setting refers to the resistance from the bottom terminal to the wiper. Of course, total potentiometer resistance is measured from the top terminal to the bottom terminal.

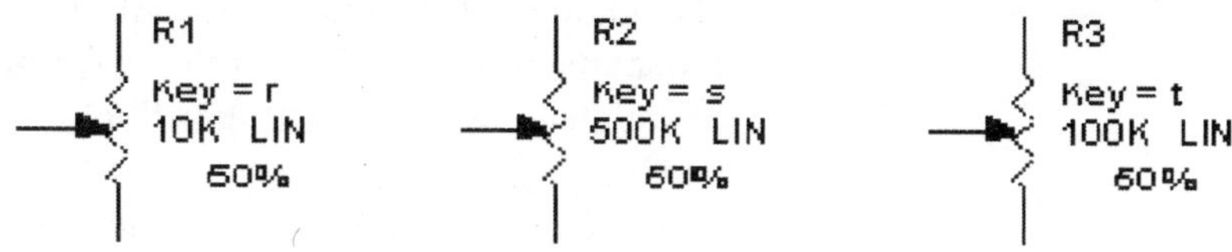


**Figure 2-10** Potentiometers in *Electronics Workbench®*

3. Open circuit file **02-07a**. Use the DMM to measure total resistance of each potentiometer and enter the data in Table 2-5. Notice that the DMM is connected and ready to measure the total (top to bottom) resistance of $R_1$.

| | Total Resistance of Potentiometer | Resistance at Starting Setting | Resistance at 25% Setting | Resistance at 75% Setting |
|---|---|---|---|---|
| $R_1$ [R] | | | | |
| $R_2$ [S] | | | | |
| $R_3$ [T] | | | | |
| $R_4$ [U] | | | | |
| $R_5$ [V] | | | | |

**Table 2-5** Varying Potentiometer Resistance

4. Measure the resistance of each potentiometer from the bottom terminal to the wiper and enter the results in the "Resistance at Starting Setting" column in Table 2-5. Then adjust each potentiometer with the proper keyboard key for a reading at the 25% point and the 75% point. Record the results in Table 2-5.

5. Next, open circuit file **02-07b**. Activate the circuit and record the ammeter and voltmeter readings with $R_2$ set for 100%.

6. Change the potentiometer ($R_2$) settings to 75%, 50%, and 25%. Record the total current in the circuit ($I_T$) and the voltage drop across $R_2$ (Voltmeter1) for each setting in Table 2-6. This project demonstrates that the voltage drop across the potentiometer is proportional to the portion of total circuit resistance that the potentiometer represents. For example, if the potentiometer is 50 Ω at a 50% setting, then it only represents one-third of the total circuit resistance (150 Ω) and will only drop one-third of the applied voltage.

| | Ammeter 1 Reading ($I_T$) | Voltmeter 1 Reading ($V_{R2}$) |
|---|---|---|
| $R_2$ at 100% | | |
| $R_2$ at 75% | | |
| $R_2$ at 50% | | |
| $R_2$ at 25% | | |

**Table 2-6** Varying Voltage and Current with a Potentiometer

Start

## Activity 2.8: Validating Ohm's Law

1. Georg Simon Ohm formulated the relationship between voltage, current, and resistance in 1827. His contemporaries laughed so much at his newly-stated formula that Ohm resigned his professorship in Germany and departed. Soon, the validity of Ohm's formula was found to be accurate and Ohm returned to his teaching. His law states that the amount of current that flows in an electrical circuit is directly proportional to the amount of voltage applied to the circuit and indirectly proportional to the resistance of the circuit. In other words, the more voltage there is, the more current flow there is in the circuit; and the more resistance there is to oppose the current flow, the less current flow there is in the circuit. The relationship can be shown mathematically as: $I = V/R$.

2. The first step is to validate the concept of the "directly proportional" relationship between current and voltage. Open circuit file **02-08a**. In this experiment, the resistance of $R_1$ is going to remain at 1 kΩ. Only the voltage will be changed to observe the resulting change in circuit current.

3. Change the value of the voltage applied to the circuit from 2 V to 4 V, 6 V, 8 V, 10 V, and 12 V, and in turn, record the corresponding change in the value of $I_T$, in Table 2-7.

| | $V_A = 2$ V | $V_A = 4$ V | $V_A = 6$ V | $V_A = 8$ V | $V_A = 10$ V | $V_A = 12$ V |
|---|---|---|---|---|---|---|
| $I_T =$ | 2mA | 4mA | 6mA | 8mA | .01 A | .012 A |

**Table 2-7** Proportionality of Voltage and Current

4. Notice that the current increased proportionally with the increase in voltage. What would happen if the voltage were changed to 20 V? Calculated $I_T$ = 20 mA. Change the voltage to 20 V and verify your calculation. Measured $I_T$ = 20 mA. or ,020 A

5. The second step is to validate the "inversely proportional" relationship between current and resistance. Open circuit file **02-08b**. The voltage of the voltage source $V_1$ is going to remain at 10 V throughout this experiment.

6. Change the value of the resistance ($R_1$) from 1 kΩ, to 2 kΩ, to 5 kΩ, to 10 kΩ and finally to 20 kΩ. Record in Table 2-8, the resulting changes in $I_T$ as the resistance values are changed.

| | $R_1 = 1$ kΩ | $R_1 = 2$ kΩ | $R_1 = 5$ kΩ | $R_1 = 10$ kΩ | $R_1 = 20$ kΩ |
|---|---|---|---|---|---|
| $I_T =$ | ,0,1 A | 4.9mA | 2.0mA | .99mA | .49mA |

**Table 2-8** Inverse Relationship of Resistance and Current

$I = \frac{E}{R} = \frac{10V}{500\Omega} = ,02mA$

7. Notice that the current flow decreased proportionally as the resistance increased. What would happen if the amount of resistance in this circuit were changed to 500 Ω? Calculated $I_T$ = .02 mA.

8. Now that the relationships between current, voltage, and resistance are understood, it is possible to calculate any one of the three variables simply by knowing the value of the other two variables.

$I \quad \frac{25}{5,000}$

9. Open circuit file **02-08c**. Notice that a different key on the keyboard activates the on-off switch for each circuit. In Circuit 1, calculate the value of $I_T$ using the known values of $V_1$ and $R_1$. $I_T$ = .005 mA.

10. In Circuit 2, calculate the value of $V_2$ using the known values of $R_1$ and $I_T$.

    $V_2$ = 150 V.

11. In Circuit 3, calculate the value of $R_3$ using the known values of $V_3$ and $I_T$.

    $R_3$ = .25 kΩ.

- ***Troubleshooting Problem:***

12. Open circuit file **02-08d**. There is something wrong with this circuit. According to Ohm's law calculations, the value of total current is incorrect. Use the DMM and measure $V_1$ and $R_1$ to determine the problem(s).

    The problem(s) is/are Batt Labeled wrong (Act 40V) Resistor labeled wrong.

## Activity 2.9: Calculating Power Consumption

1. **Power** can be defined as the rate of using energy and **energy** is the ability to do work. **Electrical energy** is the ability to do electrical work. The rate of expenditure of that electrical energy is measured in a unit called the **watt.** The basic formula for power is "Power (in watts) is equal to voltage (in volts) times current (in amperes)" or $P = V \times I$.

2. Open circuit file **02-09a**. Calculate how much power the circuit is consuming. $P_T$ = 40 W. (Use the $P = V \times I$ formula.)

3. By combining the power formula with the Ohm's law formula, other mathematical methods of determining power consumption are possible. The three formulas are: $P = I \times V$, $P = V_2/R$, and $P = I_2 \times R$.

4. Open circuit file **02-09b**. Determine how much power the circuit is consuming. $P_T$ = 2.4 W. (Use the $P = V_2/R$ formula.)

5. Open circuit file **02-09c**. How much power is the circuit consuming?

   $P_T$ = 180 W. (Use the $P = I_2 \times R$ formula.)

- ***Troubleshooting Problem:***

6. Open circuit file **02-09d**. There is something wrong with this circuit. The circuit should consume 720 W of power and according to the stated voltage and the current reading, the circuit is only consuming 360 W of power. Use the DMM and measure $V_1$ and $R_1$ to determine the problem.

   The problem is Voltage labeled wrong (act 60V). Resistor is labeled correct

# 3. Electric Circuits

**_References_**
*Electronics Workbench®, MultiSIM* Version 6
*Electronics Workbench®, MultiSIM* Version 6 Study Guide

**Objectives** After completing this chapter, you should be able to:

- Determine the requirements for a complete circuit.
- Recognize a series circuit configuration.
- Recognize a parallel circuit configuration.
- Recognize a series-parallel circuit configuration.
- Determine current paths in a complex circuit.
- Measure voltage and current in a circuit.
- Troubleshoot a circuit using voltage and current meters.

## Introduction

One of the job requirements for electrical/electronic technicians is to be able to troubleshoot electronics equipment by comparing the electrical parameters of a circuit under test with the schematic diagram of the circuit. The technician has to be able to interpret the diagram and break the circuit down into discrete segments that are recognizable as series, parallel, series-parallel, and even more complex combinations of components. The schematics of electrical circuits are usually much more complex than the circuits that are studied in the classroom and constructed in the school laboratory. During the troubleshooting process, it is necessary to break these complex circuits down into basic building blocks, and then to isolate the problem. The building block concept is used in electronics troubleshooting by means of the block diagram, a type of drawing. These blocks are the various combinations of components that make the circuit work.

Knowledge concerning electrical circuits in general and the ability to properly use electronics test equipment to determine the operation of a circuit under test is a normal job requirement for all electrical/electronic technicians. Using the voltmeter, ammeter, ohmmeter, and other pieces of test equipment to troubleshoot can be easily learned in the school setting. The student can make use of EWB to build a virtual circuit on the computer monitor and then compare the virtual information with a "real" circuit built on the test bench in the laboratory.

## Activity 3.1: Recognizing a Complete Circuit

1. As has been previously stated, the basic circuit consists of a voltage/current source, a load, a control device, and the connecting means. In the circuit of Figure 3-1, the voltage source is a battery power supply ($V_1$), the load is the lamp ($U_1$), the control device is the switch ($S_1$), and the connectors are the wires between the components. On some electronics schematics, the designator, **L** for lamp, is used.

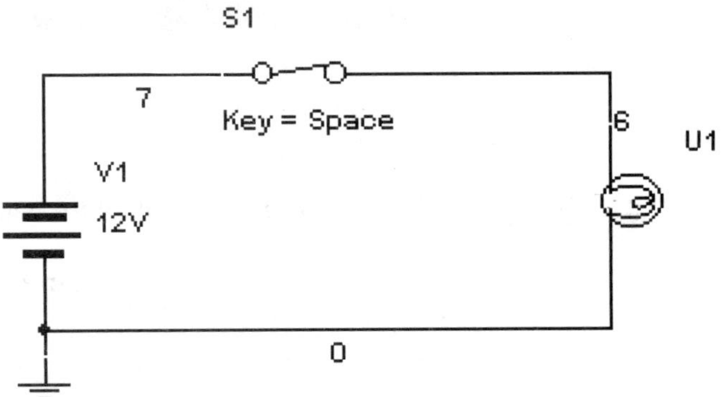


**Figure 3-1** A Basic Circuit in EWB

2. Open circuit file **03-01**. This circuit contains all of the components that are necessary to meet the requirements of a basic circuit. In addition, there is an ammeter installed in the circuit to indicate total circuit current flow. Activate the circuit and actuate Switch 1 so that current flows in the circuit. What is the total amount of current ($I_T$) flowing in the circuit as indicated by meter $M_1$? $I_T$ = 833.3 mA.

3. Disconnect the wire between TPA and TPB. What is the amount of current ($I_T$) flowing in the circuit now? $I_T$ = 0.0 A.

4. When you see a current reading in EWB, with the pico (p) symbol as part of the value indicated on the meter, you usually can interpret this reading as a zero reading or something very close to that value. This is especially true when a much larger value is expected. This generality has to be taken with a degree of caution because some readings will normally be in the picovolt or picoamp region and the p symbol is to be expected in those cases. You, as a technician, should always have a rough approximation of what kind of voltage or current level to expect before making a measurement.

5. With the connection removed between TPA and TPB, is the circuit complete?

   Yes ______ or No X.

## Activity 3.2: Recognizing a Series Circuit

1. A series circuit can be recognized as a circuit that has **one** complete path for current to flow through. Open circuit file **03-02**. This is a series circuit with one path for current flow.

2. Activate the circuit. Close switch $S_1$ to complete the circuit. Does the lamp glow? Turn the circuit off (make sure that the circuit is deactivated before installing the ammeter). This is a very important practice for the test instruments that you use on the test bench; turn off power before connecting.

3. Insert the ammeter into the circuit between the switch and the lamp by dragging the ammeter onto the wire between the components. Reactivate the circuit and determine the value of $I_T$ as indicated by the ammeter.

   Measured $I_T$ = 833 mA.

## Activity 3.3: Recognizing a Parallel Circuit

1. A parallel circuit can be recognized as a circuit with more than one path for current to flow through. Open circuit file **03-03**. This is a parallel circuit with two lamps as a parallel load. Notice that there are two paths for current flow, one path through Lamp 1 ($U_1$) and the other path through Lamp 2 ($U_2$).

2. Activate the circuit. Close switches $S_1$ and $S_2$ to complete the circuit. Do both lamps glow? Yes X or No ____. Turn the circuit off (make sure that the circuit is deactivated before installing the ammeter).

3. Insert the ammeter into the circuit between the switch and the lamps. Reactivate the circuit. What is the value of $I_T$ as indicated by the ammeter?

   Measured $I_T$ = 1667 mA.

4. Disconnect Lamp 2 ($U_2$) from the circuit by opening $S_2$. Reactivate the circuit. Does Lamp 1 ($U_1$) still glow? Yes X or No ____. What happened to the current as indicated by the ammeter? $I_T$ = 833 mA.

## Activity 3.4: Recognizing a Series-Parallel Circuit

1. Series-parallel circuits have two primary criteria that make them series-parallel circuits. First, the circuit has one or more components that are in series with the power source (in other words, all current flows through

them) and secondly, there are also components that are in parallel with other components. In the parallel portions of this type of circuit, the current divides between the parallel branches. Open circuit file **03-04**. This is a simple series-parallel circuit with three lamps, $U_1$ and $U_2$ in parallel with each other and in series with $U_3$ and all three lamps in series with the power source.

2. Activate the circuit. Close all of the switches to complete the circuit. Do all of the lamps glow? Yes _____ or No _____. What is the value of $I_T$ as indicated by the ammeter? Measured $I_T$ = __________ mA. Turn the circuit off.

3. Disconnect Lamp 2 ($U_2$) from the circuit by opening $S_2$. Reactivate the circuit and determine which lamps still glow. $U_1$ _____, $U_2$ _____, and/or $U_4$ _____? What happened to current as indicated by the ammeter? $I_T$ = __________ mA. Reconnect Lamp 2 ($U_2$) back into the circuit by closing $S_2$. Activate the circuit and make sure that the current has returned to the original value that was indicated by the ammeter in Step 2.

4. Disconnect Lamp 3 ($U_4$) from the circuit at by opening $S_3$. What happened to current flow? $I_T$ = __________ mA. Does this prove that Lamp 3 ($U_3$) is in series with the rest of the circuit? Yes _____ or No _____. Reconnect Lamp 3 ($U_4$) back into the circuit by closing $S_3$.

## Activity 3.5: Determining Current Paths in a Complex DC Circuit

1. A complex circuit may consist of various series, parallel and series-parallel combinations of components which can be reduced to simple values, yet this type of circuit cannot be reduced to a simple circuit because it contains more than one power source.

2. Open circuit file **03-05a**. This is a complex circuit with multiple paths for current flow. In this circuit, resistors are being used as loads with ammeters in each of the possible paths for current flow. All of the resistors are 1 kΩ and that makes the circuit easy to evaluate.

3. Did you notice how the current that flows into a node is equal to the current flowing out of the node? This action agrees with the concepts of Kirchhoff's Current Law. The algebraic sum of the currents flowing in and out of a circuit node is equal to zero; or in other words, what flows in, flows out.

4. Activate the circuit and then close switch $S_1$ to complete the circuit. Enter the readings from the ammeters in Table 3-1.

| | $M_1$ | $M_2$ | $M_3$ | $M_4$ | $M_5$ | $M_6$ |
|---|---|---|---|---|---|---|
| Readings in mA | | | | | | |

**Table 3-1** Circuit Ammeter Readings

5. If any one of the individual components is disconnected (open), then the current flow for the entire circuit changes. If the open component is a series component there will be no current flow in the circuit. If the open component is a parallel component, then current will continue to flow through the other parallel component(s) and the series circuit current will change according to the amount of influence the open (parallel) component has on the overall circuit.

- ***Troubleshooting Problems:***

6. Open circuit file **03-05b**. In this circuit, which is the same as the previous circuit of Steps 1 through 3, there are two open components. By comparing the current readings in this circuit with the ones from the previously investigated circuit of Steps 1 through 3, you should be able to locate which resistors are open. Which resistors are open? The open resistors

   are R _____ and R _____. What is the value of total current ($I_T$)?

   Measured $I_T$ = __________ mA. In this circuit, both $M_1$ and $M_6$ are in the series current path and both display the same value ($I_T$) of current.

7. The operation of the circuit has changed from its normal condition because of the two open components. What is the best definition for this circuit configuration with the two open components? The best definition for this circuit in its present condition is that it is a series / parallel / series-parallel circuit (circle the best answer).

## Activity 3.6: Measuring Voltage in a Circuit

1. Voltage is measured in a circuit by connecting the test leads of the test equipment across the component or at test points in a circuit where you desire information. In this activity, the same circuit that was used in Activity 3.5 will be used to measure voltage drops across individual components.

2. Open circuit file **03-06a**. In this circuit, $M_1$ measures $I_T$ and the DMM will be used to measure individual voltage drops across each of the resistors. Activate the circuit and notice that the DMM is already connected across $R_1$. What is the value of the voltage drop across $R_1$ ($V_{R1}$)?

   $V_{R1}$ = __________ V.

3. Notice that the test leads from the DMM are connected so that the positive terminal is connected to the more positive end of the resistor (concerning voltage potential) and the negative terminal is connected to the more negative end. The negative end of any component is always closest in voltage potential to the negative terminal on the power source (marked in this circuit by the negative symbol at the bottom of the power source).

4. Calculate the current through each resistor, measure the voltage drops across the resistors and record the data in Table 3-2. What is the value of

   $I_T$? $I_T$ = __________ mA. If the test leads are reversed when they are connected to the circuit, the display will indicate a negative answer.

| | $R_1$ | $R_2$ | $R_3$ | $R_4$ | $R_5$ | $R_6$ |
|---|---|---|---|---|---|---|
| Current Calculations | | | | | | |
| Readings in Volts | | | | | | |

**Table 3-2** Voltage Readings in a Circuit

5. Open circuit file **03-06b**. Use the DMM to determine voltage drops and current flow through each of the components in this circuit. Record the data in Table 3-3.

| | $R_1$ | $R_2$ | $R_3$ | $R_4$ | $R_5$ | $R_6$ | $R_7$ |
|---|---|---|---|---|---|---|---|
| Measured Current | | | | | | | |
| Voltage Readings | | | | | | | |

**Table 3-3** Circuit Current and Voltage Data

## • *Troubleshooting Problems:*

6. Open circuit file **03-06c**. In this circuit there are two open components. Activate the circuit and notice the current flow. The fact that there is current flow indicates that the open components are parallel components and not series components.

7. When resistors are in parallel and, seemingly, one of them is open, it is difficult to determine which one is open without visual observation (burnt?) or by disconnecting them one at a time and observing the resultant action on the voltage and current meters. You could also insert a current meter in each parallel branch to check for proper current flow in that branch. In this circuit, when an open resistor is disconnected, there will be no change in the voltage or current reading. When a "non-open" resistor is disconnected, there will be definite changes in the circuit operation as observed on the meters.

8. Disconnect resistors and determine which ones are open? The open resistors are R____ and R____. What is the value of total current ($I_T$) in this circuit? $I_T$ = __________ mA.

9. What would happen to this circuit if $R_6$ burned open? If $R_6$ burned open, ______________________________. What would happen to total current ($I_T$) in the circuit? $I_T$ = __________ mA.

# 4. Analyzing and Troubleshooting Series Circuits

***References***
*Electronics Workbench®, MultiSIM* Version 6
*Electronics Workbench®, MultiSIM* Version 6 Study Guide

**Objectives** After completing this chapter, you should be able to:

- Recognize a series circuit.
- Measure and record current and voltage measurements in a series circuit.
- Determine voltage drops in a series circuit.
- Prove the accuracy of Kirchhoff's voltage law.
- Analyze a series circuit to determine total resistance.
- Analyze series circuits to determine voltage, resistance, current, and power requirements of all components in the circuit.
- Prove the concept of circuit ground (common).
- Recognize voltage readings as being related to circuit ground or as voltage drops.
- Analyze series aiding and opposing voltage sources.
- Analyze unloaded voltage dividers.
- Troubleshoot series circuits.

## Introduction

As has been previously stated, series circuits are circuits in which the same amount of current flows through every component in the circuit. As a technician, you have to understand the characteristics of a series circuit to be able to interpret schematic diagrams and diagnose circuit faults.

In a series circuit, any open component in the circuit will prevent current flow. If one of the components changes to a lower value or shorts (zero resistance), then the current flow will increase. At the same time, if one of the components increases in resistance, then the current flow will decrease. With these concepts in mind, it can be stated that resistance and current flow in a series circuit are inversely proportional.

When using schematic diagrams, it is necessary to interpret the various voltage readings on the schematic and compare them with the actual voltage found in the circuit. Many schematic diagrams have expected circuit voltages printed on the schematic at key locations to help service technicians troubleshoot defective equipment. There are some companies that produce schematics with test information and key voltages and waveforms to help the technician in the troubleshooting process.

It is imperative that the technician be able to understand and properly use electronics test equipment to determine the proper operation of the circuit under test and to determine which component(s) are defective.

## Activity 4.1: Recognizing a Series Circuit

1. In review, a series circuit consists of a voltage source, a load, a control device, and the necessary connections. The components are connected in such a way that all of the circuit current flows through all of the components.

2. Open circuit file **04-01**. There are three circuits shown; determine which circuit is a series circuit. Circuit __________ is a series circuit. In this circuit, all of the current flows through all of the components.

3. An ammeter is connected in Circuit 2 to measure ($I_T$). Current flow is always through an ammeter and it has to be connected in series with the rest of the circuit. In this circuit, $I_T$ = __________ mA.

4. Use the DMM as a voltmeter to measure the voltage drops in Circuit 2 across the three resistors and record them in Table 4-1.

| | $V_{R4}$ | $V_{R5}$ | $V_{R6}$ |
|---|---|---|---|
| Resistor Voltage Drops | | | |

**Table 4-1** Voltage Drops in a Series Circuit

## Activity 4.2: Measuring Voltage and Current in a Series Circuit

1. Open circuit file **04-02a**. This series circuit has a voltmeter connected across each of the resistors. Activate the circuit and record the voltage drops for $V_{R1}$, $V_{R2}$, and $V_{R3}$ in Table 4-2.

| | $V_{R1}$ | $V_{R2}$ | $V_{R3}$ |
|---|---|---|---|
| Resistor Voltage Drops | | | |

**Table 4-2** More Voltage Drops in a Series Circuit

2. Insert the ammeter into the circuit to measure $I_T$ between TPA and TPB.

   Measured $I_T$ = ___________ mA.

3. What is the voltage output of the voltage source?

   Displayed $V_A$ = ___________ V.

4. Open circuit file **04-02b**. Connect the voltmeters across the three resistors to measure the voltage drops. Activate the circuit and record the voltage drops for $V_{R1}$, $V_{R2}$, and $V_{R3}$ in Table 4-3. The positive end of the meter is always connected to the most positive end of the resistor which is the end closest to the positive terminal of the power supply.

| | $V_{R1}$ | $V_{R2}$ | $V_{R3}$ |
|---|---|---|---|
| Resistor Voltage Drops | | | |

**Table 4-3** Voltage Drops in a Series Circuit

5. Insert the ammeter into the circuit to measure $I_T$. $I_T$ = ___________ mA.

6. There is an obvious relationship between the voltage drops across the resistors and the amount of resistance. Using the Ohm's law formula $V = I \times R$, it is possible to calculate the individual voltage drops across the individual resistors. According to Kirchhoff's voltage law for series circuits, the sum of the individual voltage drops is equal to the voltage applied ($V_A$) to the circuit by the power source.

## • *Troubleshooting Problems:*

7. Open resistors prevent current from flowing in a series circuit. An open resistor has infinite resistance and all of the applied voltage will be dropped across the open. In this case, the other resistors in such a circuit will measure a 0 V voltage drop (or at least, a very small voltage drop) across each of them. The open resistor in a series circuit is the largest resistance (infinite resistance) in the series path and the other resistors are small in comparison.

8. Open circuit file **04-02c**. Use the DMM to measure the voltage drops across each of the resistors. Activate the circuit and record the voltage drops across the resistors. Enter the data, $V_{R1}$, $V_{R2}$, and $V_{R3}$ in Table 4-4.

| | $V_{R1}$ | $V_{R2}$ | $V_{R3}$ |
|---|---|---|---|
| Resistor Voltage Drops | | | |

**Table 4-4** More Voltage Drops in a Series Circuit

9. Which resistor is open? R __________ is open.

## Activity 4.3: Using Kirchhoff's Voltage Law in a Series Circuit

1. Kirchhoff's law states that "the sum of the voltage drops in a series circuit is equal to the voltage applied to the circuit" or "the sum of the voltage drops is equal to the voltage rise." Each voltage drop has negative to positive polarity from one end of the load to the other. One end of the load is closer to one power source terminal than the other. There is a negative to positive voltage drop and the end of the load (for example) closest to the negative terminal of the power source is more negative than the other end which is closer to the positive terminal.

2. Open circuit file **04-03a**. Measure the voltage at Nodes TPA, TPB, TPC, and TPD in reference to ground and record them in Table 4-5. In reference to ground means that the negative terminal on the DMM is connected to ground for the purpose of this exercise.

| | TPA | TPB | TPC | TPD |
|---|---|---|---|---|
| Voltage Measurements | | | | |

**Table 4-5** Kirchhoff's Voltage Law in a Series Circuit

3. Notice that the voltage, starting at TPD and moving toward TPA, is progressively more positive (in reference to ground). The individual negative to positive voltage drops across each load add up to the applied voltage.

4. If the voltage at TPE were measured, what would it be?

   TPE = __________ V.

- ***Troubleshooting Problem:***

5. Open circuit file **04-03b**. Activate the circuit and notice that there is a problem; the current flow is too high. The problem is R ____, which is ______________________________.

## Activity 4.4: Determining Total Resistance in a Series Circuit

1. Total resistance in a series circuit is equal to the sum of the individual resistances. You simply add them together.

2. Open circuit file **04-04**. Calculate total resistance ($R_T$) in the circuit.

   $R_T$ = __________ Ω. Use the DMM to verify the calculation.

## Activity 4.5: Determining Unknown Parameters in Series Circuits

1. If any two factors in a circuit such as voltage, resistance, current, or power consumption are known about a particular component, then the remaining (unknown) factor(s) about that component can be determined. In a series circuit, the most important parameter to know is current flow, because current flow is the same everywhere in the circuit. Use Ohm's law, Kirchhoff's voltage law, and known characteristics about series circuits to solve for unknown factors in this activity.

2. Open circuit file **04-05a**. In this circuit, the known parameters for each of the resistors are current flow and voltage drop. Solve for the resistance of the three resistors. Calculated values are: $R_1$ = ____________ Ω,

   $R_2$ = ____________ Ω, and $R_3$ = ____________ Ω.

3. Open circuit file **04-05b**. Solve for current in this circuit using $V_A$ and $R_T$. $I_T$ = __________ μA or __________ mA. Use the DMM to verify your calculations.

4. Open circuit file **04-05c**. Solve for the value of applied voltage in this circuit using Ohm's law, $V_A = I_T \times R_T$. $V_A$ = ____________ V.

5. Open circuit file **04-05d**. Solve for voltage in this circuit using Kirchhoff's voltage law, $V_A = V_1 + V_2 + V_3$. $V_A$ = __________ V.

6. Open circuit file **04-05e**. Activate the circuit and solve for unknowns, filling in the empty blanks in the following partial solution matrix (Table 4-6). The known factors about the circuit are already entered in the table.

| Component | Resistance | Voltage | Current |
|---|---|---|---|
| $R_1$ | | 4.4 V | |
| $R_2$ | 4.7 kΩ | | |
| $R_3$ | | 13.6 V | |
| $R_4$ | 10 kΩ | | |
| $R_5$ | | 6.6 V | |
| Total | | | |

**Table 4-6** Determining Series Circuit Parameters

7. Calculate the power requirements for each of the resistors in Table 4-6.

   $P_{R1}$ = ____ W, $P_{R2}$ = ____ W, $P_{R3}$ = ____ W, $P_{R4}$ = ____ W, and

   $P_{R5}$ = ____ W. What is the amount of total power consumed by the circuit?

   $P_T$ = ____ W.

## ● *Troubleshooting Problems:*

8. Open circuit file **04-05f**. Using the solution matrix of the previous exercise and the DMM, find the faulty resistor in this circuit and describe the fault. The faulty resistor is R ____ because ________________________.

9. Open circuit file **04-05g**. Using the solution matrix of Step 6 and the DMM, find the faulty resistor in this circuit and describe the fault. The faulty resistor is R ____ because ________________________.

10. Open circuit file **04-05h**. Using the solution matrix of Step 6 and the DMM, find the faulty component in this circuit and describe the fault. The faulty component is R ____ because ________________________.

11. Open circuit file **04-05i**. Using the solution matrix of Step 6 and the DMM, find the faulty component in this circuit and describe the fault. The faulty component is R ____ because ________________________.

## Activity 4.6: Grounding the Circuit

1. In all circuits there is a point of reference that is considered to be 0 V. This point is referred to as circuit **ground** or **common**. Usually this point in the circuit is the negative terminal of the power source. Ground in a circuit is wherever the designer determines that ground needs to be located. This is the point of reference when voltages are being measured. This excludes voltage drop measurements where a voltage measurement is made across a component or between two points in the circuit rather than from ground. When measuring a voltage, decide whether the measurement is in reference to ground, which requires the connection of the negative terminal of the DMM to ground, or if a voltage drop measurement is being made where the test leads are connected across the device being measured.

2. Open circuit file **04-06a**. Ground is connected to the negative terminal of the power source (normal connection). Measure the voltage readings at the four test points and record the data on the first line (at TPD) in Table 4-7.

| Ground Location | TPA | TPB | TPC | TPD |
|---|---|---|---|---|
| At TPD | | | | |
| At TPC | | | | |
| At TPB | | | | |
| At TPA | | | | |

**Table 4-7** Data for Variable Grounds Locations

3. Move the ground terminal from TPD to TPC. Measure the voltages again and record the data on the second line (at TPC) of Table 4-7.

4. Move the ground terminal from TPC to TPB. Measure the voltages again and record the data on the third line (at TPB) of Table 4-7.

5. Move the ground terminal from TPB to TPA. Measure the voltages again and record the data on the last line (at TPA) of Table 4-7.

6. Notice that some of the voltages become negative when the ground location is moved away from the negative terminal of the power source toward the positive terminal. In fact, with the ground at TPA, all of the readings are negative in respect to ground. Return the ground terminal connection to TPD where it was originally located.

7. When referring to voltage drops such as the voltage drop across $R_2$ (TPB to TPC), the voltage would be referred to as $V_{R2}$ or $V_{BC}$. The first reference letter of $V_{BC}$ (the $_B$) indicates the measurement point and the second letter (the $_C$) indicates the reference point. Measure voltage $V_{AC}$.

   $V_{AC}$ = ________ V.

8. Usually, $V_A$ indicates the voltage applied to the circuit and not the voltage at TPA in reference to ground. If there were a TPA in the circuit, then $V_A$ would be the same as TPA. What is the voltage at TPC ($V_C$)?

   Measured $V_C$ = ________ V.

## • *Troubleshooting Problems:*

9. Open circuit file **04-06b**. Activate the circuit and close the switch. Now smoke is pouring out of the power source and the ammeter reads 12 GAmps. What is wrong? The problem is ____________________

   ________________________________________.

10. Normally, a fuse would protect a power source of this type or there would be internal circuitry that would prevent such a massive overload. Notice that the current being drawn from the power source is 12,000,000,000 amperes according to $M_1$.

# Activity 4.7: Series-Aiding and Series-Opposing Power Sources

1. In electronics equipment, such as personal computers, many different power sources are needed to operate various types of circuits with differing voltage and current requirements. Sometimes the needed voltages are positive and sometimes negative. In some configurations the power sources aid one another and in other configurations the power sources oppose one another.

2. When power sources aid one another, the output voltages add. Open circuit file **04-07a**. What do the voltages being displayed by $M_1$, $M_2$, and $M_3$ represent ($M_1$ is given as an example)? $M_1$ displays 5 V representing the V1 voltage, $M_2$ displays ____________________, and $M_3$ displays ____________________.

3. When power sources oppose one another, the output voltages algebraically add. Open circuit file **04-07b**. What do the three voltages displayed on $M_1$, $M_2$, and $M_3$ represent?

   $M_1$ displays ________________________________________,

   $M_2$ displays ________________________________________, and

   $M_3$ displays ________________________________________.

4. Open circuit file **04-07c**. What is the algebraic sum of the four voltage sources ($V_{AE}$)? Calculated $V_{AE}$ = ____________. Use the DMM to verify your calculation. Measured $V_{AE}$ = ____________.

5. Determine the value of TPA ($V_A$), TPB ($V_B$), TPC ($V_C$), TPD ($V_D$), and TPE ($V_E$). $V_A$ = ____________ V, $V_B$ = ____________ V, $V_C$ = ____________, $V_D$ = ____________, and $V_E$ = ____________ V.

6. What is amount of current flowing through $R_1$? $I_{R1}$ = ____________ mA. Verify the calculation with the DMM.

## Activity 4.8: Unloaded Voltage Dividers and Resistor Power Requirements

1. A voltage divider is a series of resistors that are connected in an arrangement that provides specific voltages at junctions between the resistors. A voltage divider is analyzed by determining the voltages at the junctions between the resistors in reference to ground.

2. Open circuit file **04-08**. Measure the voltages at TPA ($V_A$), TPB ($V_B$), and TPC ($V_C$). Measured $V_A$ = ____________ V, $V_B$ = ____________ V, and $V_C$ = ____________ V. How much current is flowing in the circuit? $I_T$ = ____________ mA.

3. What are the power requirements of the resistors in this circuit? $P_{R1}$ = ____________ W, $P_{R2}$ = ____________ W, $P_{R3}$ = ____________ W, and $P_{R4}$ = ____________ W.

4. What is the total power consumed by the circuit? $P_T$ = ____________ W.

# 5. Analyzing and Troubleshooting Parallel Circuits

**References**

*Electronics Workbench®, MultiSIM* Version 6

*Electronics Workbench®, MultiSIM* Version 6 Study Guide

**Objectives** After completing this chapter, you should be able to:

- Recognize a parallel circuit.
- Measure and record current and voltage measurements in a parallel circuit.
- Determine current paths in a parallel circuit.
- Prove the accuracy of Kirchhoff's current law.
- Analyze a parallel circuit to determine total resistance.
- Analyze a parallel circuit to determine voltage, current, resistance, and power requirements of all components in the circuit.
- Troubleshoot parallel circuits.

## Introduction

Parallel circuits are circuits in which the voltage is the same across all of the parallel components and the current divides between the parallel branches. Parallel circuits are commonly found in residential wiring where most of the AC loads (outlets) in a residence are in parallel. Another example is in automotive circuits where all of the electrical devices are in parallel across the DC power source. It is necessary for technicians to understand the characteristics of parallel circuits to enable them to interpret schematic diagrams and diagnose circuit faults in these types of circuits.

In a parallel circuit, a component can be open in one parallel branch and the current will still flow through the other parallel branches. If one of the parallel components changes to a lower value of resistance or shorts out (zero resistance), then the current flow will increase through that component alone and may cause an overloading condition for the power source, which could result in a tripped circuit breaker or a blown fuse. If one of the parallel components increases in resistance, then the current flow will decrease through that component and not affect the other parallel branches. An open parallel branch will have little effect other than to reduce the total current drain on the power source as load decreases.

## Activity 5.1: Recognizing a Parallel Circuit

1. In a manner similar to series circuits, a parallel circuit consists of a voltage source, two or more loads in parallel, a control device, and the necessary connections. All parallel components are connected in such a way that the source current divides between the parallel branches. A parallel circuit can be thought of as being similar to a river in which the current divides to go around an island and then joins again after passing the island.

2. Open circuit file **05-01**. How many parallel branches are there in the circuit?

   There are __________ branches in the circuit.

3. In this circuit, the meters are connected in a manner that demonstrates the path of current flow, leaving the power source, dividing between the branches (going around the islands), and then rejoining before proceeding on to the other power source terminal. The $M_1$ and $M_2$ meter readings represent total current and the $M_3$, $M_4$, and $M_5$ meter readings represent the branch currents.

4. Activate the circuit and enter the current flow data in Table 5-1. You can see that the current flowing into the branches is equal to the current flowing out from the branches and that total current is equal to the sum of the branch currents. What is the total current $I_T$? $I_T$ = __________ mA.

| | $I_{Branch1}$ | $I_{Branch2}$ | $I_{Branch3}$ |
|---|---|---|---|
| Resistor Currents in Each Branch | | | |

**Table 5-1** Branch Current Data

5. Place $R_4$ in parallel with the circuit by connecting it to TPA and TPB. What happened to $I_T$? $I_T$ changed to __________ mA. How much current is flowing through $R_4$? $I_{R4}$ = __________ mA.

6. Whenever another resistor is placed in parallel with the rest of the circuit, the circuit current $I_T$, increases / decreases (circle the best answer).

## Activity 5.2: Measuring Voltage and Current in a Parallel Circuit

1. The current in each parallel branch is equal to the voltage across that branch divided by the resistance of that branch ($I_{Rx} = V_{Rx} \times R_x$).

2. Open circuit file **05-02a**. This parallel circuit has a voltmeter connected across the circuit. Activate the circuit, and calculate and record the current through each parallel path in Table 5-2.

| | $I_{R1}$ | $I_{R2}$ | $I_{R3}$ |
|---|---|---|---|
| Resistor Currents in Each Branch | | | |

**Table 5-2** Branch Current Measurements

3. Measure $I_T$. Measured $I_T$ = __________ mA.

4. Determine the voltage output of the voltage source with the DMM.

   Measured $V_A$ = __________ V. Measure the voltage across each of the resistors. Measured $V_{R1}$ = __________ V, $V_{R2}$ = __________ V, and $V_{R3}$ = __________ V.

5. There is an obvious relationship between the current flow in each branch of a parallel circuit and the amount of resistance in each branch. Using Ohm's law, it is possible to calculate the individual currents flowing in each branch of the circuit. Another observation is that the sum of the branch currents in a parallel circuit is equal to the total current of the circuit ($I_T$).

6. Using Ohm's law, determine total resistance $R_T$. $R_T$ = __________ Ω.

7. Open resistors in a parallel branch prevent current from flowing in that particular branch of the circuit. The total current flowing in the parallel circuit will be less than expected by the amount that should be flowing in the faulty current path (branch). Applied voltage will be dropped across the faulty component, but there will be no current flow. Other branches in the circuit will provide paths for current flow and the total current will be the sum of the conducting paths. When determining current flow for a parallel circuit, and the total current is less than expected (according to the calculations), then one of the branches is likely to be open.

### ● *Troubleshooting Problems:*

8. Open circuit file **05-02b**. Calculate the expected total current for the circuit.

   Calculated $I_T$ should be __________ mA.

9. Activate the circuit. $I_T$ is less than it should be.

   Measured $I_T$ = __________ mA. Use the DMM to measure the current in each branch of the circuit. Notice that the DMM is in position to measure the current in the $R_1$ branch. Use this as a model and record the branch currents in Table 5-3.

| | $I_{R1}$ | $I_{R2}$ | $I_{R3}$ |
|---|---|---|---|
| Branch currents | | | |

**Table 5-3** Branch Currents in a Parallel Circuit

10. Which resistor is open? R __________ is open. What should be the amount of current flowing in the open branch? The current in the open branch should be __________ mA.

11. Open circuit file **05-02c**. Calculate the value of total current. $I_T$ should be __________ A. What value of current should be flowing through the $R_3$ branch of the circuit. $I_{R3}$ should be __________ A. But, $I_{R3}$ = __________ A. What do you think is wrong? The problem is that $R_3$ has a value of __________ Ω and should be __________ Ω.

## Activity 5.3: Using Kirchhoff's Current Law in a Parallel Circuit

1. Kirchhoff's law states that the algebraic sum of the currents flowing in and out of a junction (node) is equal to zero. In simpler terms, the amount of current flowing into a junction is equal to the amount of current flowing out of the junction.

2. Open circuit file **05-03a**. The node equation for this circuit in reference to TPA (Node A) is $I_T = I_{R1} + I_{R2} + I_{R3}$. Calculate the circuit current flow and the branch currents and rewrite the equation using the displayed current values. The node equation for TPA using calculated current values would be ______________________________.

3. Activate the circuit and use the DMM to measure the currents in each of the parallel branches and record the data in Table 5-4. Also record $I_T$.

| | $I_T$ | $I_{R1}$ | $I_{R2}$ | $I_{R3}$ |
|---|---|---|---|---|
| Current Measurements | | | | |

**Table 5-4** Using Kirchhoff's Current Law

- ***Troubleshooting Problems:***

4. Open circuit file **05-03b**. Activate the circuit and notice that there is a problem in this circuit. $I_T$ is too low and there is little or no current flowing through $R_2$ and $R_3$. Use the DMM to isolate the problem. The problem is ______________________________.
Remember that the problem does not have to be a component; it can be a wiring problem.

5. Try to bypass the problem. Connect a jumper wire from the bottom of TPB to the bottom of TPD and see if the circuit works.

## Activity 5.4: Determining Total Resistance in a Parallel Circuit

1. Ohm's law is the easiest method to determine total resistance in a parallel circuit and should be used whenever possible.

2. Open circuit file **05-04a**. Activate the circuit and notice the voltage across the circuit and the current through the circuit. Calculate total resistance ($R_T$) in the circuit using Ohm's law. Using Ohm's law, $R_T$ = __________ Ω. Open Switch $S_1$ and connect the DMM across the parallel circuit and verify the calculation.

   The DMM measures __________ Ω.

3. Another method of determining total resistance of a parallel circuit with many branches is by using the **reciprocal method**. This method is valuable when voltage and current parameters are unavailable or difficult to obtain and it is necessary to know the amount of load a circuit offers to a power source.

4. Open circuit file **05-04b**. Calculate the total resistance of Circuit 1 and Circuit 2 using the reciprocal formula. Calculated $R_T$ for

   Circuit 1 = __________ kΩ.

   Calculated $R_T$ for Circuit 2 = __________ Ω.

5. Verify the calculations with the DMM. Measured $R_T$ for

   Circuit 1 = __________ kΩ.

   Measured $R_T$ for Circuit 2 = __________ Ω.

6. Another method of determining total resistance of a parallel circuit is by using the **conductance method**. Essentially, the conductance method is the same as the reciprocal method because $G = 1/R_T$.

7. Open circuit file **05-04c**. Calculate the total resistance of Circuit 1 and Circuit 2 using the conductance formula.

   Calculated G for Circuit 1 = __________ Ω.

   Calculated G for Circuit 2 = __________ Ω.

   Calculated $R_T$ for Circuit 1 = __________ Ω.

   Calculated $R_T$ for Circuit 2 = __________ Ω.

8. Verify the calculations with the DMM.

   Measured $R_T$ for Circuit 1 = __________ Ω.

   Measured $R_T$ for Circuit 2 = __________ Ω.

9. Another method of determining total resistance of a parallel circuit is by using the **product-over-sum method**, $R_T = (R_1 \times R_2)/(R_1 + R_2)$.

10. Open circuit file **05-04d**. Calculate the total resistance of Circuit 1 and Circuit 2 using the product-over-sum method formula.

    Calculated $R_T$ for Circuit 1 = __________ Ω.

    Calculated $R_T$ for Circuit 2 = __________ kΩ.

11. Verify the calculations with the DMM. Measured $R_T$ for

    Circuit 1 = __________ Ω. Measured $R_T$ for Circuit 2 = __________ kΩ.

12. Another method of determining total resistance of a parallel circuit is by using the **equal-value-resistor method**, $R_T = R/N$.

13. Open circuit file **05-04e**. Calculate the total resistance of Circuit 1 and Circuit 2 using the equal-value-resistor method.

    Calculated $R_T$ for Circuit 1 = __________ kΩ.

    Calculated $R_T$ for Circuit 2 = __________ Ω.

14. Verify the calculations with the DMM.

    Measured $R_T$ for Circuit 1 = __________ kΩ.

    Measured $R_T$ for Circuit 2 = __________ Ω.

15. In Circuit 2, if another 80 Ω resistor were added in parallel, what would $R_T$ equal? $R_T$ = __________ Ω.

16. A final method of determining total resistance is by using the **assumed voltage** method.

17. Open circuit file **05-04f**. Calculate $R_T$ using the assumed voltage method.

    Calculated $R_T$ = __________ Ω. Measure $R_T$ with the DMM.

    Measured $R_T$ = __________ Ω.

18. Activate the circuit, and using $R_T$ and $I_T$ according to $M_1$, calculate $V_A$.

    What is the value of $I_T$? $I_T$ = __________ A. $V_A$ = __________ V.

19. What is the total power ($P_T$) consumed by this circuit? Use any one of the power formulas to calculate circuit power consumption.

    $P_T$ = __________ W.

## • *Troubleshooting Problems:*

20. Open circuit file **05-04g**. Calculate $R_T$ and then $I_T$ according to the schematic. Calculated $R_T$ = __________ Ω and $I_T$ = __________ mA.

21. What does the current meter indicate the circuit $I_T$ is at present? Circuit $I_T$ = __________ mA and thus, $R_T$ = __________ Ω. What is wrong with the circuit? The problem with the circuit is ______________________.

22. Obviously there are some problems with the resistors. Replace the defective resistors with spares located to the right of the circuit and then recheck the circuit. $I_T$ = __________ mA when the circuit is repaired.

## Activity 5.5: Determining Unknown Parameters in a Parallel Circuit

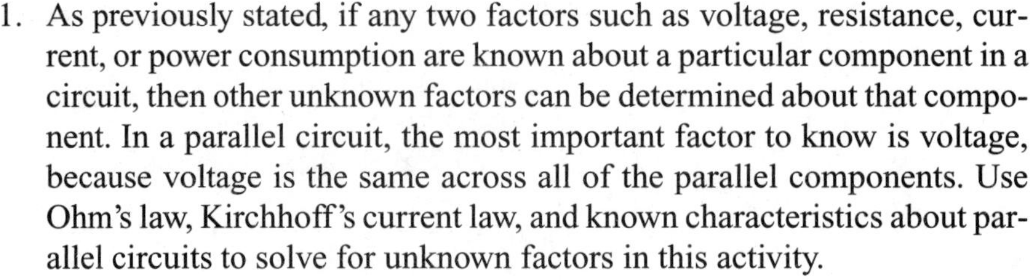

1. As previously stated, if any two factors such as voltage, resistance, current, or power consumption are known about a particular component in a circuit, then other unknown factors can be determined about that component. In a parallel circuit, the most important factor to know is voltage, because voltage is the same across all of the parallel components. Use Ohm's law, Kirchhoff's current law, and known characteristics about parallel circuits to solve for unknown factors in this activity.

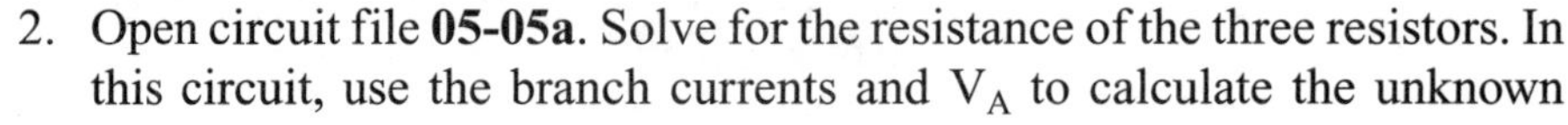

2. Open circuit file **05-05a**. Solve for the resistance of the three resistors. In this circuit, use the branch currents and $V_A$ to calculate the unknown resistors. $R_1$ = __________ Ω, $R_2$ = __________ Ω, and $R_3$ = __________ Ω.

   Determine the value of $R_T$. $R_T$ = __________ Ω.

3. What is $P_T$ and what is the power consumption of each of the resistors?

   $P_T$ = __________ mW, $P_{R1}$ = __________ mW, $P_{R2}$ = __________ mW, and $P_{R3}$ = __________ mW.

4. Open circuit file **05-05b**. Solve for $I_T$ first, then for $R_T$, and then the resistance of each of the other resistors. $I_T$ = __________ A.

   $R_T$ = __________ Ω, $R_1$ = __________ Ω, $R_2$ = __________ Ω, and $R_3$ = __________ Ω.

5. Open circuit file **05-05c**. Calculate the voltage drop across each of the resistors. $V_{R1}$ = __________ V, $V_{R2}$ = __________ V, and $V_{R3}$ = __________ V.

   What is the value of $V_A$? $V_A$ = __________ V.

6. What are the values of $I_T$ and $R_T$ in this circuit? $I_T$ = __________ mA and $R_T$ = __________ Ω. Verify the calculations with the DMM.

7. Open circuit file **05-05d**. Solve for current in this circuit using Kirchhoff's current law, $I_T = I_{R1} + I_{R2} + I_{R3}$. $I_T$ = __________ mA.

8. What is the value of $R_T$ in this circuit? $R_T$ = __________ Ω.

9. The next circuit will be a mixture of all of the techniques that you have learned, up to this point, to be able to solve for unknowns in partially specified parallel circuits.

10. Open circuit file **05-05e**. Measure and calculate to solve for unknowns and fill in the empty blanks in the following partial solution matrix (Table 5-5). Verify the calculations with the DMM.

| Component | Resistance | Voltage | Current |
|---|---|---|---|
| $R_1$ | | | 1.667 mA |
| $R_2$ | 12 kΩ | | |
| $R_3$ | 21 kΩ | | 3.571 mA |
| $R_4$ | 15 kΩ | | |
| $R_5$ | | | 7.5 mA |
| Totals | | | |

**Table 5-5** Determining Parallel Circuit Parameters

11. Calculate the power requirements for each of the resistors in Table 5-5.

$P_{R1}$ = ________ W, $P_{R2}$ = ________ W, $P_{R3}$ = ________ W,

$P_{R4}$ = ________ W, and $P_{R5}$ = ________ W.

What is total power consumed by the circuit? $P_T$ = ________ W.

## • *Troubleshooting Problems:*

12. Open circuit file **05-05f**. Using the solution matrix of the previous exercise and the DMM, find the faulty resistor in this circuit and describe its fault. The faulty resistor is R ________ because ________________

________________. What does $I_T$ measure? $I_T$ = ________ mA.

Is it right? Yes ____ No ____.

13. Open circuit file **05-05g**. Using the results determined in the solution matrix of Step 10 and the DMM, find the faulty resistor in this circuit and describe its fault. The faulty resistor is R ________ because

________________________________.

What does $I_T$ measure? Measured $I_T$ = ________ mA.

Is the measurement right? Yes ____ No ____.

# 6. Analyzing and Troubleshooting Series-Parallel Circuits

***References***

*Electronics Workbench®, MultiSIM* Version 6

*Electronics Workbench®, MultiSIM* Version 6 Study Guide

**Objectives** After completing this chapter, you should be able to:

- Recognize a series-parallel circuit.
- Simplify a series-parallel circuit.
- Analyze a series-parallel circuit to determine voltage, current, resistance, and power requirements of all components in the circuit.
- Determine characteristics of unloaded and loaded voltage divider circuits.
- Observe effects and causes of voltmeter loading.
- Recognize and analyze bridge circuits.
- Troubleshoot series-parallel circuits.

## Introduction

**Series-parallel circuits** are circuits in which the characteristics of series and parallel circuits are combined. It is necessary to use all of the techniques learned up to this point to determine the operational parameters of these more difficult circuits. In electronics equipment, series-parallel circuits are used more often than the other types of circuits that have been studied thus far. When working on equipment it is unusual to find simple series or parallel circuits in the equipment and it is more common to spend the bulk of the time working on very complex series-parallel combinations. Technicians need to fully understand the characteristics of series-parallel circuits to enable them to interpret schematic diagrams and diagnose circuit faults in electronics equipment.

A series-parallel resistive circuit can be reduced to a single resistance for the purpose of determining total circuit load offered by the circuit. It is possible to construct a particular series-parallel circuit in EWB and determine the operational characteristics of that circuit on the computer monitor without ever touching the actual circuit. In this chapter, you will use virtual instrumentation to verify circuit parameters.

## Activity 6.1: Recognizing a Series-Parallel Circuit

1. As a combination of series and parallel circuits, a series-parallel circuit has the characteristics of both circuits and has to be isolated into discrete sections. Within this type of circuit one portion of the circuit can be a parallel section and another portion can be a series section. It is necessary to take into consideration all of the circuit rules that you have learned up to this point for these two types of circuits.

2. First, determine which components are in series in the circuit. Open circuit file **06-01a**. How many series resistors are there in the circuit? There are __________ series resistors in the circuit. How many resistors are in parallel in the circuit? There are __________ parallel resistors in the circuit.

3. Open circuit file **06-01b**. This has a number of series and parallel sections. Identify them. How many resistors are in series? There are __________ resistors in series in this circuit. How many parallel sections are there in the circuit? There are __________ parallel sections in the circuit.

4. In this circuit, there is one section with three resistors in parallel. The three resistors in parallel are R__________, R__________, and R__________. Total resistance would be calculated as $R_T = R_1 + R_2 + (R_3 \parallel R_4) + R_5 + R_6 + R_7 + (R_8 \parallel R_9) + (R_{10} \parallel R_{11}) + R_{12} + R_{13} + (R_{14} \parallel R_{15}) + R_{16} + (R_{17} \parallel R_{18} \parallel R_{19})$. In this text, the ‖ symbol means "in parallel with."

   Calculated $R_T =$ __________ Ω. Use the DMM and measure the total resistance of the circuit. Measured $R_T =$ __________ Ω.

5. There are also circuits where resistors are in series within a parallel section. Open circuit file **06-01c**. Total resistance in this circuit would be calculated as $R_T = R_1 + (R_2 \parallel [R_3 + R_4]) + R_5 + (R_6 \parallel [R_7 + R_8]) + R_9 + R_{10} + (R_{11} \parallel R_{12} \parallel [R_{13} + R_{14}]) + R_{15}$. Calculated $R_T =$ __________ Ω. Use the DMM and measure the total resistance of the circuit.

   Measured $R_T =$ __________ Ω.

## Activity 6.2: Simplifying Series-Parallel Circuits

1. The secret to simplifying series-parallel circuits is by treating each section as an individual circuit, simplifying the individual sections by reducing them to a single resistance, and then adding up all of the series components along with the reduced (simplified) sections. Generally, the best procedure to follow is to work from the point farthest from the power source, simplifying each section as you work toward the power source. It is possible to end up with the sum of all the series components and simplified sections becoming a single resistance value that offers a single load to the power source.

2. Open circuit file **06-02a**. This circuit has one parallel section consisting of $R_3$, $R_4$, and $R_5$. To simplify the circuit, calculate the resistance of the parallel section (call the equivalent resistance of that section "$R_A$").

   $R_A$ = ________ Ω. To obtain a value for $R_T$, add the resistance of the reduced section to the three series resistors.

   $R_T = R_1 + R_2 + R_A + R_6 =$ ________ Ω.

3. Open circuit file **06-02b**. Simplify this circuit. $R_T$ = ________ Ω. The formula would be $R_T = R_1 + R_2 + [R_3 \parallel R_4 \parallel R_6] + [R_6 \parallel R_7] + [R_8 \parallel R_9]$.

## Activity 6.3: Measuring Voltage and Current in a Series-Parallel Circuit

1. Open circuit file **06-03a**. Activate the circuit and notice that the current meters demonstrate that the current leaves the power source, divides between the branches (going around the islands), and then rejoins before proceeding on to the other power source terminal.

2. Activate the circuit and observe that the $M_1$ and $M_2$ meter readings represent total current. In this circuit, $I_T$ = ________ mA. The $M_3$, $M_4$, and $M_5$ meter readings represent the branch currents. The total of the branch currents, $I_{R1}$ = ________ mA, $I_{R2}$ = ________ mA, and $I_{R3}$ = ________ mA, is equal to the current flowing into the branches. In other words, $I_{R1} + I_{R2} + I_{R3}$ = ________ mA $= I_T$. The current flowing into the branches is equal to the current flowing out from the branches. $M_6$ represents the voltage drops across $R_3$, $R_4$, and $R_5$ (the parallel section).

3. Take a good look at the current flow in the circuit. Starting from the negative terminal of the power source, $I_T$ flows through resistor R ________ and then splits up to flow through resistors R ________, R ________, and

R ________, rejoins, and then flows through resistors R ________ and R ________.

Resistors R ________, R ________, and R ________ are in series with the circuit and all of the current flows through them.

Resistors R ________, R ________, and R ________ are in parallel with each other, but as a group are in series with the rest of the circuit. If one of the series resistors $R_1$, $R_2$, or $R_6$ were open, then there would be no current flow in the circuit because the series current path would be open.

## • *Troubleshooting Problems:*

4. Open circuit file **06-03b**. Activate the circuit and determine the problem.

   The faulty series resistor, R ____, is ________ and measures ______ Ω.

   What happened to $I_T$? $I_T$ ________________________________.

5. Open circuit file **06-03c**. Activate the circuit and determine the problem.

   The faulty parallel resistor, R ____, is ________ and measures ______ Ω.

   What happened to $I_T$? $I_T$ ________________________________.

6. Open circuit file **06-03d**. Activate the circuit and determine the problem.

   The faulty resistor, R ____, is ________ and measures ______ Ω. What happened to $I_T$? $I_T$ ________________________________.

7. Open circuit file **06-03e**. Activate the circuit and determine the problem.

   The faulty resistor, R ____, is ________ and measures ________ Ω.

   What happened to $I_T$? $I_T$ ________________________________.

8. In series-parallel circuits, you should have a rough idea of what the voltage drops are going to be and where the current is going. It is usually difficult, when troubleshooting electronics equipment, to make current measurements because it is difficult to break into the circuit to insert a meter. In some equipment, provision is made to take current measurements with special current test points. Technicians rely to a large degree on voltage drop measurements for the bulk of their troubleshooting (at least early in the course of the troubleshooting process). They place most of their emphasis on the voltage measurements and later move to somewhat more complex types of test equipment and different types of measurements.

9. Open circuit file **06-03f**. The voltage drop across the $R_3$-$R_4$ parallel circuit should measure 30 V and it measures ________ V. What is wrong with this circuit? The problem is ______________________________ ______________________________. (This circuit has two faulty resistors.)

## Activity 6.4: Loaded and Unloaded Voltage Divider Circuits

1. An unloaded **voltage divider** is a series circuit that has certain (usually according to design) voltages at the junctions between the resistors. This type of arrangement makes up a voltage divider circuit.

2. Open circuit file **06-04a**. Use the DMM and determine the voltages at TPA, TPB, and TPC. Voltages at TPA = ________ V, at TPB = ________ V, and at TPC = ________ V.

3. Open circuit file **06-04b**. This is a loaded voltage divider. Measure the voltages across the loads ($R_4$ and $R_5$). What are the load voltages? The load voltages are $V_{R4}$ = ________ V and $V_{R5}$ = ________ V. What amount of current is being drawn by the loads? The load currents are $I_{R4}$ = ________ mA and $I_{R5}$ = ________ mA.

4. Open circuit file **06-04c**. The bottom resistor ($R_3$) in a voltage divider is called a bleeder resistor. When designing a voltage divider, the first step is to determine bleeder current. This design starts with a bleeder current of 50 mA. Select the bleeder resistor first, which has a design requirement of 10 V at 50 mA according to the specifications.

   $R_3$ = ________ Ω.

5. Next, determine the value of $R_2$, which drops 20 V at a current of 60 mA (the 60 mA is the bleeder current of 50 mA plus the load current of 10 mA). $R_2$ = ________ Ω. The voltage drop of 20 V is the voltage difference between the ends of the resistor, $R_2$.

6. Lastly, the value of $R_1$ needs to be determined. $R_1$ has a current of 80 mA and drops 70 V. $R_1$ = ________ Ω. What are the wattage requirements of the three resistors $R_1$, $R_2$, and $R_3$?

   $P_{R1}$ = ________ W, $P_{R2}$ = ________ W, and $P_{R3}$ = ________ W.

### • *Troubleshooting Problems:*

7. Open circuit file **06-04d**. Yesterday this was a perfectly running voltage divider circuit, supplying the correct voltages to the loads. Something is wrong, the voltages aren't right today. Did one of the loads change or did a divider resistor change value? Measure the voltage divider resistors and the loads. The problem is ____________________________________.

8. Open circuit file **06-04e**. Now there are more problems, earlier today this was a perfectly running voltage divider circuit, supplying the correct voltages to the loads (with a bleeder current of 100 mA). Something is wrong; the 10 V load is smoking and the output voltages aren't right. Did one of the loads change or did something happen in the divider circuit?

   The problem is ____________________________________.

## Activity 6.5: The Effects of Voltmeter Loading

1. One of the types of circuits that is easiest to "load down" is the voltage divider. In this exercise a voltage divider will be used to demonstrate the loading effect that a low resistance voltmeter presents to a circuit. All voltmeters have an input resistance characteristic that sometimes causes problems when they are placed in parallel with a "touchy" load such as a precision voltage divider. Usually, the more inexpensive the meter, the more potential there is for loading problems (this is typical with the inexpensive flea market meter). And, of course, "in parallel" is the way that voltmeters are used. Most modern meters (digital types) have high input resistance ratings and do not affect a circuit to any great extent, but in the case of the "touchy" circuit or when readings that just are a "little bit off," the meter always has to be taken into consideration.

2. Open circuit file **06-05a**. This circuit is supposed to provide voltages that are approximately 10 V at TPA and 30 V at TPB. Activate the circuit and notice that the voltages are slightly incorrect. What are the readings?

   The voltage at TPA is __________ V and

   the voltage at TPB is __________ V.

3. Disconnect meters $M_1$ and $M_2$ from the circuit and measure the voltages at TPA and TPB with the DMM. Now the readings are correct. Leave the DMM hooked up to TPA and reconnect $M_1$ and $M_2$. Now, the voltages become wrong again. Obviously meters $M_1$ and $M_2$ have an affect on the operation of the voltage divider while the DMM does not seem to affect the circuit. Click on the **Settings** button on the DMM and a menu will come up to inform the viewer that the DMM resistance is 1 GΩ. Now, right-click on the panel meters $M_1$ and $M_2$. Notice that the resistance

reading is set at 100 kΩ. Change the setting to 1 MΩ and check the circuit again. What are the voltage readings now? The voltage at TPA is

__________ V and the voltage at TPB is __________ V, and that is better; but if the meters are set to a higher resistance (try 10 MΩ), the voltage divider readings will be even closer.

- ***Troubleshooting Problem:***

4. Open circuit file **06-05b**. The 10 V source is out of tolerance; what is wrong?

   The problem is ______________________________________________.

## Activity 6.6: Dealing with Bridge Circuits

1. There is a classification of series-parallel circuits known as **bridge circuits**. These types of circuits are used in electronics instrumentation to interface sensors or to facilitate the measurement of natural phenomena.

2. Open circuit file **06-06a**. This circuit is a balanced bridge circuit (*Wheatstone*). The ratio of $R_1/R_2 = R_3/R_4$ has to be maintained to keep the voltmeter reading at 0 volts. $R_4$ is the resistor in this circuit that is usually used to measure the "unknown" resistance or phenomenon under test. That is the primary usage of this type of circuit, to make these types of tests. Change resistor $R_3$ = 3 kΩ and $R_4$ = 9.9 kΩ. What happened to the voltmeter reading? The ratio $R_1/R_2 = R_3/R_4$ would still be the same and

   the voltmeter reading will remain very close to __________ V. Change $R_4$ to 10 kΩ and observe the change in the voltmeter reading. What is the

   reading now? The voltmeter reading with $R_4$ = 10 kΩ is __________ mV. This is the method used to balance bridge circuits.

3. Open circuit file **06-06b**. This circuit is a balanced bridge circuit with a 10-kΩ potentiometer in the place of $R_4$. The potentiometer is set at its 50% point and the voltmeter reading indicates an unbalanced condition. What is the "percentage" setting of the potentiometer that will restore balance to the bridge circuit? The potentiometer setting would be

   __________ % of 10 kΩ.

- ***Troubleshooting Problem:***

4. Open circuit file **06-06c**. This bridge circuit is prepared to measure an unknown resistor in the range of 0 to 50 kΩ. Notice that this circuit is using an ammeter rather than a voltmeter to measure discrete differences

between the two points. Either method will provide a zero indication when there is no potential difference between points A and B. Activate the circuit and toggle the keyboard key for the potentiometer [R] until the ammeter reads 0. At that point, the resistance of the potentiometer is equal to the unknown resistor. Multiply the percentage factor to the right of the potentiometer by the value of the potentiometer itself (50 kΩ) and that is the value of the unknown resistor. The value of the unknown resistor is $R_{unknown}$ = __________ kΩ.

# 7. Complex Circuits and Network Analysis

**_References_**

*Electronics Workbench®, MultiSIM* Version 6

*Electronics Workbench®, MultiSIM* Version 6 Study Guide

## Objectives

After completing this chapter, you should be able to:

- Identify a complex circuit.
- Analyze complex circuits.
- Simplify resistive circuits by using
  - mesh analysis,
  - nodal analysis,
  - superposition theorem techniques,
  - Thevenin's theorem,
  - Norton's theorem, and
  - pi-to-tee and tee-to-pi conversions.
- Troubleshoot complex circuits.

## Introduction

Complex circuits constitute a problem for the technician because the circuit configuration cannot be reduced to a single resistance. In the earlier circuits that we have studied (series, parallel, and series-parallel circuits), simplification techniques resulted in a single resistance being obtained. Complex circuits are categorized as complex because they contain more than one power source or they have other components that prevent the resistor networks from being simplified to one resistance. As stated previously, it is unusual to find simple series or parallel circuits in electronics equipment. Technicians spend the bulk of their time working on series-parallel combinations and other categories titled specifically as complex circuits. All technicians have to understand the characteristics of complex circuits to enable them to interpret schematic diagrams and to diagnose circuit faults in electronics equipment. The inability to reduce a complex circuit to a single resistor does not prevent the technician from diagnosing a complex circuit problem. There are many other tools, such as network theorems, that enable the circuit to be understood in its simplest terms.

## Activity 7.1: Recognizing a Complex Circuit

1. **Complex circuits** are easy to recognize because there are two or more power sources making the circuit hard to simplify. The circuit in Figure 7-1, for example, is a complex circuit. The primary factor indicating this circuit is a complex circuit is that there are two power sources.

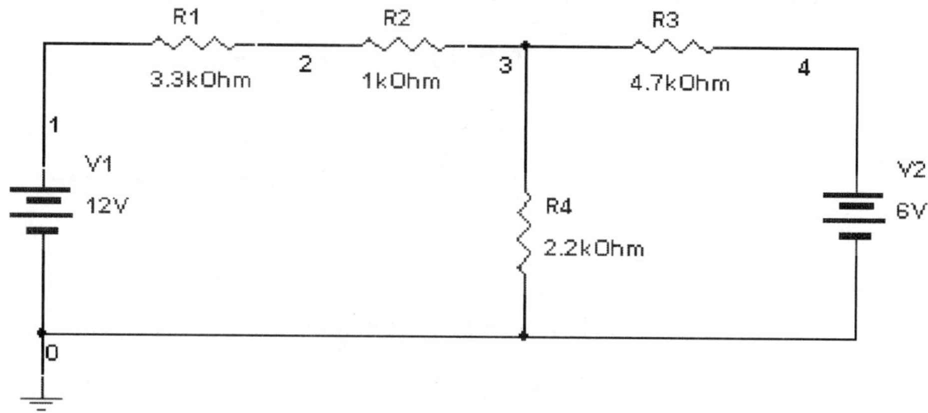


**Figure 7-1** A Complex Circuit

2. Open circuit file **07-01a**. Is it possible to simplify this circuit any further than has been accomplished on this schematic? It is possible to further simplify this circuit by ______________________________.

3. Open circuit file **07-01b**. In this complex circuit you will simplify the resistor network. Start the simplification process by combining series resistors such as $R_1$ and $R_2$ and calling the result $R_A$. $R_3$ and $R_4$ would become $R_B$ and then $R_5$ and $R_6$ would become $R_C$. What are the values of $R_A$, $R_B$, and $R_C$ after Circuit 1 is simplified? $R_A$ = __________ kΩ, $R_B$ = __________ kΩ, and $R_C$ = __________ kΩ.

4. Go to Circuit 2 and change the values of the resistors inside the $R_A$, $R_B$, and $R_C$ subcircuits to their new calculated values. Activate both circuits and verify that the two ammeters display the same value of current for both circuits. What are the two circuit currents values according to $M_1$ and $M_2$? $M_1$ = __________ mA and $M_2$ = __________ mA.

## Activity 7.2: Mesh Analysis

1. **Mesh analysis** is a mathematical method of simplifying complex circuits. This method extends the application of Kirchhoff's voltage law.

2. Open circuit file **07-02a**. In this circuit it is necessary to write mesh equations to solve circuit parameters. Write mesh equations for the circuit and solve them for Ix and It. Calculated Ix = ____________ mA and It = ____________ mA.
A typical mesh equation for this circuit in Kirchhoff's voltage law format is shown in Figure 7-2. Kirchhoff's voltage law format shows the equality between the voltage drops and the voltage source.

**Mesh 1:** $3.3Ix + (2.2Ix - 2.2It) = 12$
$5.5Ix - 2.2It = 12$

**Mesh 2:** $1.8It + (2.2It - 2.2Ix) = -8$
$4.0It - 2.2Ix = -8$

**Figure 7-2** Mesh Equations

3. These equations are the first steps of mesh analysis and they will have to be developed further to achieve the end result of determining the value of Mesh 1 (Ix) and Mesh 2 (It) currents. In this circuit, the mesh current It ends up as a negative quantity. This negative quantity indicates that the mesh current is in a direction that is opposite to the assumed direction of current for that current loop. Simply change the negative to a positive. The advantage to using EWB to solve for circuit parameters in this type of circuit is that you can check out the proper answers for your calculations with a virtual circuit. If there is a numerical discrepancy, the calculations can be compared with the EWB circuit.

4. Activate the circuit and observe $M_1$ and $M_2$ to check the validity of the mesh equations. $M_3$ displays the value of current flowing through $R_3$.

   What are the current values? $M_1$ = ____________ mA,

   $M_2$ = ____________ mA, and $M_3$ = ____________ mA.

## • *Troubleshooting Problem:*

5. Open circuit file **07-02b**. Use mesh analysis to determine the current flow through $R_1$. Calculated $I_{R1}$ = ____________ A. Now, activate the circuit and observe the value of $I_{R1}$ on meter $M_1$. $I_{R1}$ = ____________ A. Obviously, there is a problem. What is wrong? The problem is ____________

   ____________________________________________.

## Activity 7.3: Nodal Analysis

1. **Nodal analysis** is similar to mesh analysis, but it is based on Kirchhoff's current law rather than Kirchhoff's voltage law to solve complex network problems. A node is considered to be any point in a circuit where currents combine. Usually, ground is used as a reference point for solving circuit problems and making measurements. However, if ground were not present in the circuit, then an arbitrary reference point would be used for voltage measurements.

2. Open circuit file **07-03**. Write a nodal equation for this circuit and then solve for the unknown, Vx, which represents the voltage drop across $R_3$.

   Calculated Vx = __________ V. Remember that Vx = $V_{R3}$.

3. Activate the circuit and use the DMM to measure the voltage drop across

   $R_3$. Measured $V_{R3}$ = __________ V.

## Activity 7.4: The Superposition Theorem

1. The **superposition theorem** is another valuable technique for analyzing complex circuits that have more than one power source. With this method you act as if one of the power sources acting is a short, determining current values and voltage drops in reference to that one power source. Then you proceed to the other power source and act as if the first power source is a short, again determining current values and voltage drops in reference to the second power source. Then algebraically combine the results to determine circuit parameters.

2. Open circuit file **07-04a**. There are three circuits on the workspace. Circuit 1 represents the circuit that power source $V_1$ sees, Circuit 2 represents the circuit that power source $V_2$ sees, and Circuit 3 represents both circuits combined with ammeters in each current path. If you are unable to locate Circuit 3, use the scroll bar and move it up (this circuit is located on the workspace below Circuits 1 and 2).

3. Use the superposition theorem to calculate $I_{R3}$.

   Calculated $I_{R3}$ = __________ mA. Activate the circuit and verify the calculation with the $M_3$ display. Measured $I_{R3}$ = __________ mA.

4. Open circuit file **07-04b**. Use the superposition theorem to calculate circuit operation. First, consider V2 to be a short and calculate the circuit parameters. Then consider V1 to be a short and recalculate the circuit parameters. Finally, algebraically superimpose the results on each other. Use the multimeter to verify superposition results by measuring voltage

drops and branch currents, particularly $I_{R3}$.

Calculated $I_{R3}$ = __________ mA. Now, activate Circuit 1 to verify the calculations. Measured $I_{R3}$ = __________ mA.

- ***Troubleshooting Problem:***

5. Open circuit file **07-04c**. This circuit is the same circuit that you worked on in Step 4. Activate the circuit. Are the results the same as for Step 4? Obviously there is a problem. Determine what the problem is.

   The problem is ______________________________.

## Activity 7.5: Using Thevenin's Theorem

1. With **Thevenin's theorem**, simple resistive circuits or portions of a more complex circuit can be converted into a simple equivalent circuit. A typical Thevenin's theorem calculation results in a simplified series circuit with a constant voltage source identified as $V_{TH}$ and a single resistance identified as $R_{TH}$. In a circuit where only a portion of the circuit is being simplified using Thevenin's theorem, the point where the simplification is taking place is called the point of simplification. See Figure 7-3 for an example of a Thevenin's theorem simplified circuit.

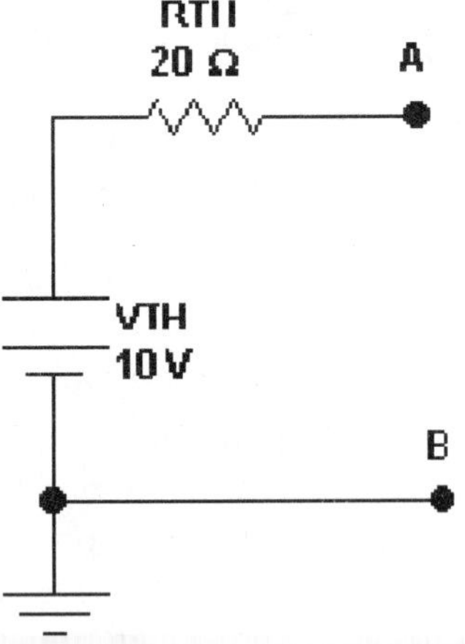


**Figure 7-3** Thevenin's Theorem Equivalent Circuit

2. As you can see in Figure 7-3, a thevenized circuit replaces a resistive circuit with a two-terminal equivalent consisting of a single voltage source called $V_{TH}$ and a single series resistance called $R_{TH}$. $V_{TH}$ is the open terminal voltage across terminals A and B. $R_{TH}$ is the single resistance in series with the voltage source and is also in series with any load attached to terminals A and B.

3. Open circuit file **07-05a**. Go through the steps involved in determining $V_{TH}$ and $R_{TH}$. Circuit A is the beginning circuit. Circuit B represents the first stage of thevenizing the circuit, obtaining the resistance of parallel components $R_2$ and $R_3$. Calculated $R_2 = R_3 =$ ________ Ω. Circuit C displays the value of $R_{TH}$ and the power source that is yet to be determined by further analysis. Circuit D displays Circuit B with $V_1 = 12$ V. Circuit E ties it all together with the calculated results of $V_{TH}$ and $R_{TH}$ being displayed. Calculated $V_{TH} =$ ________ V and $R_{TH} =$ ________ Ω.

4. Use the DMM and measure the circuit to compare the measured values of $V_{TH}$ and $R_{TH}$ with the calculated values. To obtain $V_{TH}$ and $R_{TH}$, use Circuit B for the resistance measurement and Circuit D for the voltage measurement.

   Measured $V_{TH} =$ ________ V and $R_{TH} =$ ________ Ω.

5. One important use of Thevenin's theorem is in determining how much current will flow through a load connected to output terminals A and B in the thevenized circuit.

6. Open circuit file **07-05b**. In this circuit $R_l$ is to be attached to the circuit as the load. As you can see, this is a simple series circuit and it should be easy to determine the voltage drop across the load ($R_l$) as well as the current flowing through it. Use the DMM to determine those quantities.

   $I_{LOAD} =$ ________ mA and $V_{LOAD} =$ ________ V.

● ***Troubleshooting Problem:***

7. Open circuit file **07-05c**. Calculate $V_{TH}$ and $R_{TH}$.

   Calculated $V_{TH} =$ ________ V and $R_{TH} =$ ________ Ω. Now, measure $V_{TH}$ and $R_{TH}$. Remember, $V_{TH}$ is the voltage that is measured at Test Points A and B. $R_{TH}$ is the resistance measured at Test Points A and B with the power supply shorted (place a jumper across the power supply from Test Point C to Test Point D and then disconnect the power source).

   Measured $V_{TH} =$ ________ V and $R_{TH} =$ ________ Ω. Obviously, there is a problem. What is it?

   The problem is ________________________________.

## Activity 7.6: Using Norton's Theorem

1. **Norton's theorem** is similar to Thevenin's theorem except that the final result is an equivalent circuit consisting of a current source in parallel with a resistance rather than a voltage source in series with a resistance.

Norton's theorem uses a current-related approach while Thevenin's theorem uses a voltage-related approach. A constant current source provides a current output that is unaffected by external circuit resistance. No matter what the load does, the output current remains the same. Figure 7-4 displays the EWB symbol for a constant current source and Figure 7-5 displays a Norton's theorem equivalent circuit.

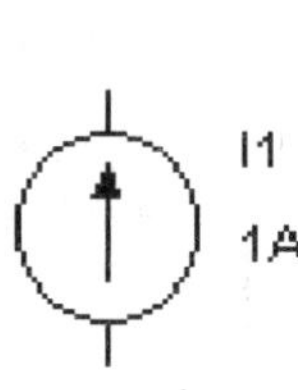


**Figure 7-4** An EWB Symbol for a Constant Current Source

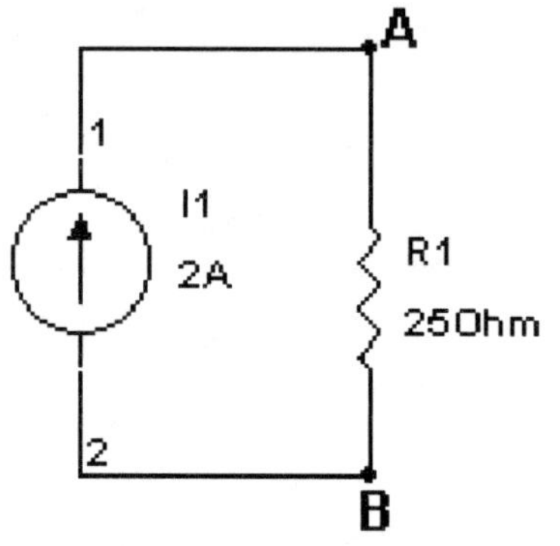

**Figure 7-5** Norton's Theorem Equivalent Circuit

2. Norton's theorem uses the same approach to calculate $R_N$ as Thevenin's theorem uses to calculate $R_{TH}$. Thus, $R_{TH}$ has the same value as $R_N$ ($R_N = R_{TH}$). The application of the calculated result is different with Norton's theorem; $R_N$ is in parallel with the current source $I_N$ and with an applied load. The Norton current divides between the load and $R_N$.

3. Open circuit file **07-06**. Circuit 1 represents the initial circuit. Start by determining the value of $R_N$. Calculated $R_N$ = __________ Ω. Circuit 2 represents the Thevenized circuit that displays the calculated value of $R_N$.

4. To calculate $I_N$, determine the amount of current that would be flowing through a shorted output from A to B which, essentially, shorts out $R_2$. This leaves $R_1$ in series with the voltage source, a simple series circuit.

   Calculated $I_N$ = __________ mA. Change the values in Circuit 4 to represent the calculated values of $R_N$ and $I_N$.

5. Connect $R_{LOAD}$ to terminals A and B of Circuit 4. Predict the current through the load and the voltage across it. The Norton's current will divide proportionally between the load resistor and $R_N$. The voltage across the load resistor (and $R_N$) will be equal to $V_{TH}$.

   Predicted $I_{LOAD}$ = __________ mA. Measure $I_{LOAD}$.

   Measured $I_{LOAD}$ = __________ mA.

## Activity 7.7: Tee-to-Pi and Pi-to-Tee Conversions

1. **Tee circuits** and **pi circuits** are also referred to in the electrical industry as **wye** and **delta circuits**. The terms **tee** and **pi** will be used in this study. See Figures 7-6 and 7-7 for examples of tee and pi circuits. The purpose of the conversion process is to attain a circuit in one of the two configurations (tee or pi) that is electrically equivalent to the other configuration. The term wye is equivalent to tee. The term delta is equivalent to pi throughout the electrical and electronics industries.

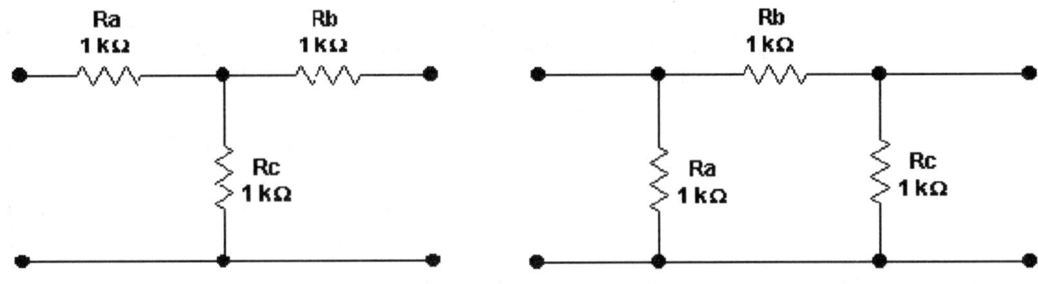


**Figure 7-6** Tee Configuration **Figure 7-7** Pi Configuration

2. The first conversion you will undertake is the tee-to-pi conversion. Open circuit file **07-07a**. In this circuit file, Circuit 1 is a tee circuit while Circuit 2 is a pi circuit. Activate the circuits and notice that the pi circuit draws almost three times more current from the source than the tee circuit and provides only about one-third more output voltage and current. Enter the data from both circuits in Table 7-1.

| | **Ra = Rb = Rc = 1 kΩ (The Original Configuration)** | | **Rab = Rbc = Rbc = ? kΩ (The Modified Configuration)** | |
|---|---|---|---|---|
| | Source 1 | Source 2 | Source 1 | Source 2 |
| Source Voltage | | | | |
| Source Current | | | | |
| | Output 1 | Output 2 | Output 1 | Output 2 |
| Output Voltage | | | | |
| Output Current | | | | |

**Table 7-1** Tee-to-Pi Conversion

3. Using the tee-to-pi formulas, change the pi circuit so that it is electrically equivalent to the tee circuit. The goal in this conversion is to find out what values in the pi circuit will provide the same load to the voltage source as the tee circuit with the same output to the load. Rab = ____ Ω, Rac = ____ Ω, and Rbc = ____ Ω. Change the values of Rab, Rac, and Rbc in Circuit 2 to match the calculations. Activate the circuit and enter the revised Circuit 2 data in Table 7-1 (the Circuit 1 data would be the same as previously entered). Notice that by changing the resistors in Circuit 2, the inputs and outputs are now identical for both circuits. That is the purpose of the conversion.

4. The second conversion you will undertake is the pi-to-tee conversion. Open circuit file **07-07b**. In this circuit file, Circuit 1 is a pi circuit and Circuit 2 is a tee circuit. Activate the circuits and notice that the tee circuit draws about one-third as much current as the pi circuit from the source and provides less output voltage and current. Enter the data from both circuits in Table 7-2.

| | **Rab = Rac = Rbc = 1 kΩ (The Original Configuration)** | | **Ra = Rb = Rc = ? kΩ (The Modified Configuration)** | |
|---|---|---|---|---|
| | Source 1 | Source 2 | Source 1 | Source 2 |
| Source Voltage | | | | |
| Source Current | | | | |
| | Output 1 | Output 2 | Output 1 | Output 2 |
| Output Voltage | | | | |
| Output Current | | | | |

**Table 7-2** Pi-to-Tee Conversion

5. Using the pi-to-tee formulas, change the tee circuit so that it is electrically equivalent to the pi circuit. The goal in this conversion is to find out what values of resistance the resistors in the tee circuit need to be able to provide the same load to the voltage source as the pi circuit and with the same output to the load. The new values are Ra = __________ Ω, Rb = __________ Ω, Rc = __________ Ω. Change the values of Ra, Rb, and Rc in Circuit 2 to match the calculations. Activate the circuits and enter the revised Circuit 2 information in Table 7-2 (the Circuit 1 data would be the same in both cases). Notice that by changing the resistors in Circuit 2, the inputs and outputs are identical for both circuits. That is the purpose of the conversion.

6. Open circuit file **07-07c**. Convert this tee circuit to an equivalent pi circuit and enter the data in Table 7-3.

   Calculated Rab = ________ Ω, Rac = ________ Ω, Rbc = ________ Ω. Change the value of the Circuit 2 components to match the converted data from Circuit 1. Activate the circuit and determine equivalency and enter your data in Table 7-3.

| | **The Original Configuration** | | **The Modified Configuration** | |
|---|---|---|---|---|
| | Source 1 | Source 2 | Source 1 | Source 2 |
| Source Voltage | | | | |
| Source Current | | | | |
| | Output 1 | Output 2 | Output 1 | Output 2 |
| Output Voltage | | | | |
| Output Current | | | | |

**Table 7-3** Tee-to-Pi Conversion Practice Problem

7. Open circuit file **07-07d**. Convert this pi circuit to an equivalent tee circuit and enter the data in Table 7-4.

   Calculated Ra = ________ Ω, Rb = ________ Ω, Rc = ________ Ω. Again note that the original configuration data for Circuit 2 has no relationship to Circuit 1 at this point. Change the resistor values of Circuit 2 to make it electrically equivalent to Circuit 1.

| | **The Original Configuration** | | **The Modified Configuration** | |
|---|---|---|---|---|
| | Source 1 | Source 2 | Source 1 | Source 2 |
| Source Voltage | | | | |
| Source Current | | | | |
| | Output 1 | Output 2 | Output 1 | Output 2 |
| Output Voltage | | | | |
| Output Current | | | | |

**Table 7-4** Pi-to-Tee Conversion Practice Problem

# 8. Electrical Power Sources

**References**

*Electronics Workbench®, MultiSIM* Version 6

*Electronics Workbench®, MultiSIM* Version 6 Study Guide

**Objectives** After completing this chapter, you should be able to:

- Connect and use power sources found in EWB *MultiSIM®*.
- Determine internal resistance of DC power sources.
- Connect DC power sources in series.
- Connect DC power sources in parallel.
- Analyze additional DC power sources found in EWB.
- Discuss and analyze AC power sources.

## Introduction

There are many methods used by industrialized nations to obtain electrical energy for industry, home, and commerce. In electronics equipment, all methods are used to one degree or another; but basically in the generation of power for electrical/electronic circuits, the primary methods of power generation are by generators (electromagnetic means) or by the use of cells and batteries (chemical means).

In *MultiSIM®*, there are several symbols that represent voltage sources as presented in Figure 8-1. Some of these symbols have already been discussed in Chapter 1, but they will be briefly covered again. Other than these few power sources in EWB, there are no other means available to power virtual circuits in the program. This excludes power generated by means of static electricity, mechanical means (piezoelectric), solar means (photovoltaic), and heat generation (thermoelectric) that are viable power sources in industry.

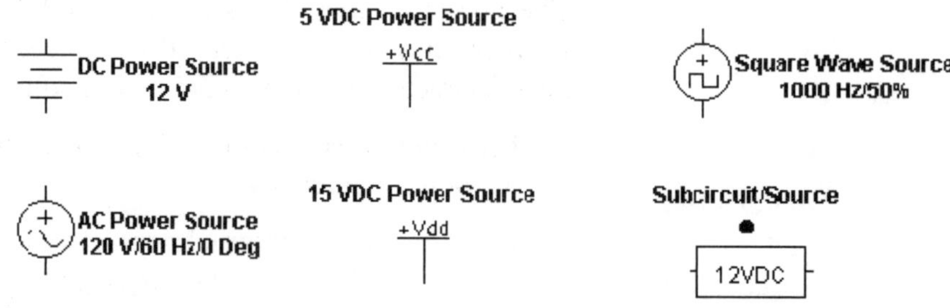


**Figure 8-1** Voltage Sources Used in *Electronics Workbench®*

Among the various DC power sources used in EWB, the source represented by a battery symbol is the source that has been used in this text up to this point (see Figure 8-2). The battery symbol is always used when there is a need for DC voltage.

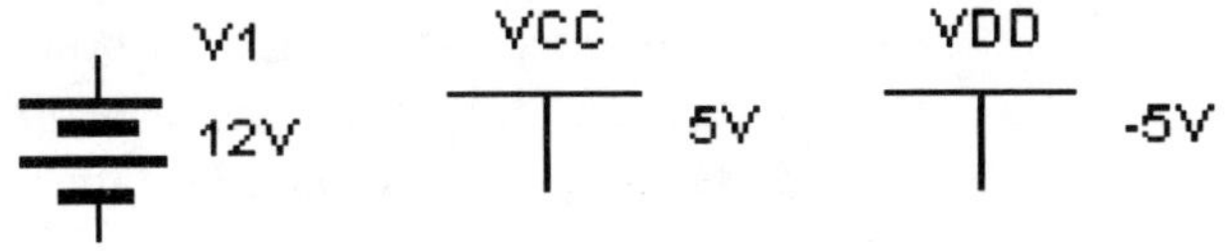

V1 12V    VCC 5V    VDD -5V

**Figure 8-2** DC Power Sources in EWB

There is a power source designated as Vxx, a voltage source where the voltage level can be set by the technician and used in many locations in a circuit. This source is primarily used for digital circuits as a +5 VDC source. It may also be used as a negative voltage source. This voltage source may be used at one voltage potential within a circuit only. It is shown in Figure 8-2 as $V_{CC}$ and $V_{DD}$.

The box with the +12 VDC caption is a subcircuit symbol. In this case, the subcircuit represents a "real" 12 VDC source with "real" internal resistance in series with the power source output. This internal resistance is a normal part of any power source. The subcircuit symbol is frequently used in EWB to place a larger circuit inside a box, with only output terminals available. This is a valuable tool for decreasing overall circuit size on the workspace and for presenting function block troubleshooting. The final EWB DC power source is the square wave signal generator, a type of pulsating voltage source, which is also variable.

Other power sources in EWB include the **alternating current** (AC) power source symbol which represents an alternating signal, as well as additional power sources with variable voltage and variable frequency capabilities. All of these AC signals can also be obtained from the function generator, which is a piece of test equipment that is kept in the **Instruments** library. After clicking on the **Instruments** library, notice that the second instrument from the left is the function generator. Drag it out and take a look at it. This instrument will be used extensively in the following chapters.

## Activity 8.1: Using DC Power Sources With Internal Resistance

1. The power sources that are used in EWB *MultiSIM*® are ideal power sources; they have no internal resistance. Ideal (virtual) power sources have characteristics that "real" power sources do not have:

   a. They do not wear out over a period of time like a battery or even an AC power source does.

   b. They do not have internal resistance, which means that they can provide a constant output voltage regardless of the load. This means that they will not blow fuses or trip circuit breakers (unless a fuse is designed into the virtual circuit) when they are overloaded.

c. They provide unlimited output current with few limitations on how many loads are connected. This is not practical; in the real world it is necessary to always be aware of how many loads a power source can handle.

2. To represent a "real" DC power source (a battery) with internal resistance, a low value resistor is placed in series with the output terminal to simulate the internal resistance of a source.

3. The subcircuit in Figure 8-1 represents a "real" DC power source, with an internal resistance in series with the power source itself.

4. Open circuit file **08-01a**. Activate the circuit and observe that the output of the power source as displayed on $M_3$ is equal to the power source voltage of 100 V. In this circuit, $R_{int}$ represents the internal resistance of the power source, a resistance of 1 Ω. Activate $S_1$ to apply a load ($R_2$) to the circuit. At this point, the load is 100 Ω and the internal resistance of the power source is only 1% of the load. This should result in about a 1% voltage drop across the internal resistance. $M_1$ represents the drop across the internal resistance, $M_2$ represents the actual voltage of the power source, and $M_3$ represents the percentage of the voltage being applied to the load.

5. Toggle the potentiometer ($R_2$) in 1% intervals and notice the drop in the voltage being applied to the load as the internal resistance drops more of the voltage. Fill in Table 8-1 with the test data.

6. The voltage across the load will equal one-half of the rated output of the power source (100 V in this case) when the potentiometer is equal to the internal resistance. What is the percentage of the potentiometer when meter $M_3$ = 50 V? The potentiometer is at __________% when $M_3$ is at 50 V. This indicates that $R_{int}$ = _____ Ω.

## *Troubleshooting Problem:*

7. Open circuit file **08-01b**. This power source has a problem; it has too much internal resistance. Determine the value of the internal resistance.

   $R_{int}$ = __________ Ω.

| Load Condition | M1 Reading | M2 Reading | M3 Reading |
|---|---|---|---|
| % and Ω | $R_{internal}$ V | Source V | (Output Voltage) |
| No Load | 0 V | 100 V | 100 V |
| 100%—100Ω | | | |
| 90%—90Ω | | | |
| 75%—75Ω | | | |
| 50%—50Ω | | | |
| 25%—25Ω | | | |
| 10%—10Ω | | | |
| 5%—5Ω | | | |
| 4%—4Ω | | | |
| 3%—3Ω | | | |
| 2%—2Ω | | | |
| 1%—1Ω | | | |
| 0%—0Ω | | | |

**Table 8-1** Power Source Loading Data

## Activity 8.2: DC Power Sources in Series

1. The voltage output of DC power sources that are placed in a series configuration always provides a voltage that is the algebraic sum of the power sources. This includes power sources that aid as well as oppose one another. In EWB, it is easy to change the voltage output of a power source. If more than one specific voltage is needed, then it is possible to stack power sources in series with voltage outputs between the individual power sources or use independent sources with different voltage outputs.

2. Open circuit file **08-02a**. This file has several circuits with various combinations of stacked power sources. What is the calculated voltage output for each of them? Use the DMM to measure the voltage outputs and then record the data in Table 8-2.

| Circuit Number | Calculated DC Output | Measured DC Output |
|---|---|---|
| Circuit 1 | | |
| Circuit 2 | | |
| Circuit 3 | | |
| Circuit 4 | | |

**Table 8-2** DC Power Sources: Series-Aiding and Series-Opposing

- ***Troubleshooting Problems:***

3. Open circuit files **08-02b**, **08-02c**, **08-02d**, and **08-02e** in turn. There is a problem with each of these power supply circuits. Calculate the voltage outputs for each circuit and then measure the outputs. Try to determine what is wrong with each power supply circuit.

4. The individual problem with each of the power supply circuits is:

   08-02b ______________________________________.

   08-02c ______________________________________.

   08-02d ______________________________________.

   08-02e ______________________________________.

## Activity 8.3: DC Power Sources in Parallel

1. When "real" DC power sources are placed in parallel (with equal voltage outputs), the voltage output is fixed. The current output is theoretically equal to the sum of the individual currents of the parallel power sources. There are problems with this type of hookup and it is not generally recommended, but can be done if necessary. The reason generally given for such a maneuver would be to provide more current than an individual source is able to provide. This type of solution, the parallel connection of power sources, is often seen in automotive applications where more than one battery is needed to provide power for auxiliary equipment. This hookup is more complicated than just connecting the batteries in parallel.

2. There should be no need to use parallel power sources in EWB because the virtual power sources have no current limitations. In fact, EWB will give an alarm condition if an attempt is made to parallel power sources.

3. Open circuit file **08-03**. Activate the circuit and an EWB alarm condition will be triggered by this method of power supply connection. EWB power sources can be connected in parallel only if a resistor is placed in series with each of the outputs. Place the 1-Ω resistors in series with the power sources and reactivate the circuit. All should be well now.

4. With the resistors still in the circuit, change the voltage setting of power source $V_1$ to 6 V. Activate the circuit and the output voltage displayed on the meter changes to 9 V. This is halfway between 12 V and 6 V; each of the resistors has a voltage drop. Use the DMM to measure the voltage drops across the 1-Ω resistors.

   Measured $V_{R2}$ = __________ V and $V_{R3}$ = __________ V.

## Activity 8.4: Additional DC Power Sources in EWB

1. This activity introduces the additional EWB DC power sources. As previously stated, the sources are **+$V_{CC}$ = +5 VDC** and **+$V_{DD}$ = +15 VDC**.

2. Open circuit file **08-04a**. Measure the +$V_{CC}$ source.

   Measured +$V_{CC}$ = ________ V.

3. Open circuit file **08-04b**. Measure the +$V_{DD}$ source.

   Measured +$V_{DD}$ = ________ V.

4. There is one more source in EWB that, technically, is a DC source. It is the square wave generator. This generator puts out a signal that rises and falls from 0 V and never goes below the 0 V point. Although the voltage acts like AC, it is truly DC. This generator will be used in a later chapter.

## Activity 8.5: Introducing the Alternating Current (AC) Power Source

1. Open circuit file **08-05**. The DMM is set up to measure the AC output of the AC power source. Activate the circuit and determine the output of the power source at point A. Notice that the DMM is set for AC rather than DC. The AC symbol is to the left of the straight line on the faceplate of the DMM.

   The AC power source output = __________ VAC.

# 9. Measuring Direct Current Parameters

**References**

*Electronics Workbench®, MultiSIM* Version 6
*Electronics Workbench®, MultiSIM* Version 6 Study Guide

**Objectives** After completing this chapter, you should be able to:

- Connect voltmeters, ammeters, and ohmmeters in DC circuits.
- Determine and adjust the internal resistance of current and voltage meters found in EWB *MultiSIM®*.
- Understand the purpose of controls/connectors/software settings for EWB test instruments.
- Calculate the value of shunt resistors for ammeters.
- Calculate the value of multiplier resistors for voltmeters.
- Use the oscilloscope as a DC measuring instrument.

## Introduction

Voltmeters, ohmmeters, ammeters, and oscilloscopes are the eyes and ears of the electronics technician whenever he/she attempts to work on the many types of electronics equipment needing service today. Whatever can be measured in electronics becomes part of the technical knowledge base that is so helpful and necessary concerning the physical operation of electronics equipment under repair or in need of adjustment.

In this study, because of the lack of hands-on capabilities, it is not possible to study electrical/electronic instruments or meters of the analog type, which have a moving-coil movement. These analog types of meters have to be physically present to be able to appreciate their characteristics. Also, they are not as prevalent as they were in the past. The type of meter that will be studied is digital and, in the case of EWB, is a virtual instrument and easy to simulate in a software program.

The technician not only has to know how to use the electronics test equipment available, but also whether a specific reading obtained with a particular piece of test equipment can be trusted. If the technician jumps to a wrong conclusion as the result of erroneous readings and/or faulty test equipment, he/she can waste a lot of time instead of promptly completing the task at hand.

## Activity 9.1: Using the Digital Multimeter (DMM) in DC Circuits

1. There are two types of voltmeters present in EWB *MultiSIM®*, the Digital Multimeter (DMM) from the **Instruments** menu and the panel type of digital meter found in the **Indicators** menu. In this activity there will be further study of the DC applications of the DMM, which have already been used extensively in previous activities. The face of the EWB multimeter is shown in Figure 9-1. The various measurement options will be discussed in the following activities.

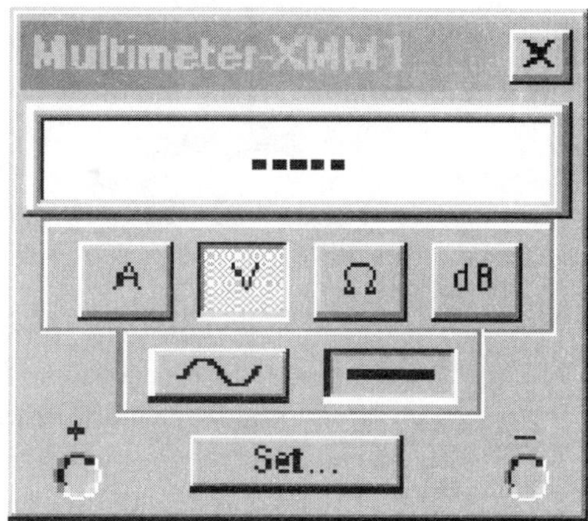


**Figure 9-1** The EWB Digital Multimeter (DMM)

2. When using the ammeter (current flow) function of the DMM, it is necessary to activate the "**A**" and the proper signal mode (DC or AC). When connecting the DMM into the circuit to measure current flow, use the two connection points at the bottom of the meter, marked "+" and "–". These connection points have polarity and have to be connected correctly when measuring quantities requiring positive and negative orientation (DC voltages and currents). For nonpolarized measurements such as resistance and AC quantities, the negative and positive orientation does not matter.

3. Open circuit file **09-01a**. In this circuit, the DMM is connected into the circuit to measure DC current. Notice the progression of current flow from the positive terminal of the power source into one end of the resistor, out of the other end of the resistor into the positive terminal of the meter, out the negative terminal of the meter, and then to the negative terminal of the power source, which completes the series circuit path. The insertion of the meter into the current flow path of the circuit requires correct polarity to get the correct negative or positive sign display on the meter face as shown in Figure 9-2. Activate the circuit and determine the

   current reading. The measured DC current reading is __________ mA.

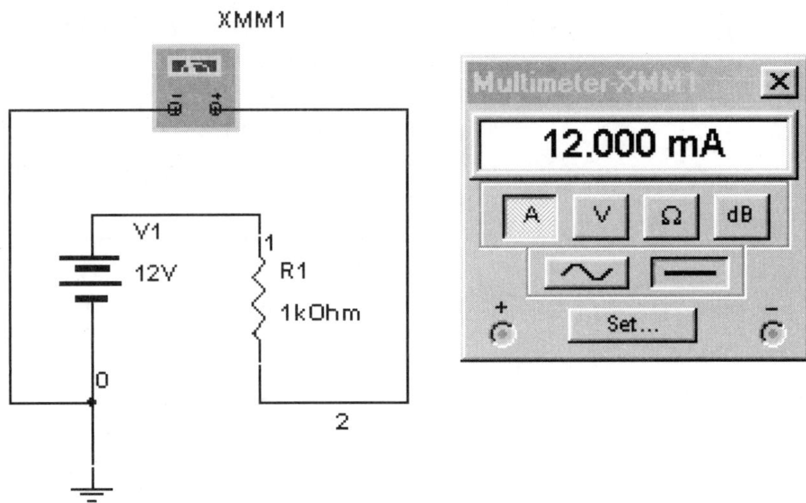


**Figure 9-2** Digital Multimeter Connection to Measure DC Current

4. The DC voltmeter function of the digital multimeter will be discussed next. Open circuit file **09-01b**. This circuit is connected to measure DC voltage. Notice the meter is now connected in parallel with the component to be measured, the resistor $R_1$. Current flows through the component and the voltmeter displays the amount of voltage the resistor is dropping in this circuit (See Figure 9-3). Activate the circuit and determine the voltage reading.

   The measured DC voltage reading is __________ V.

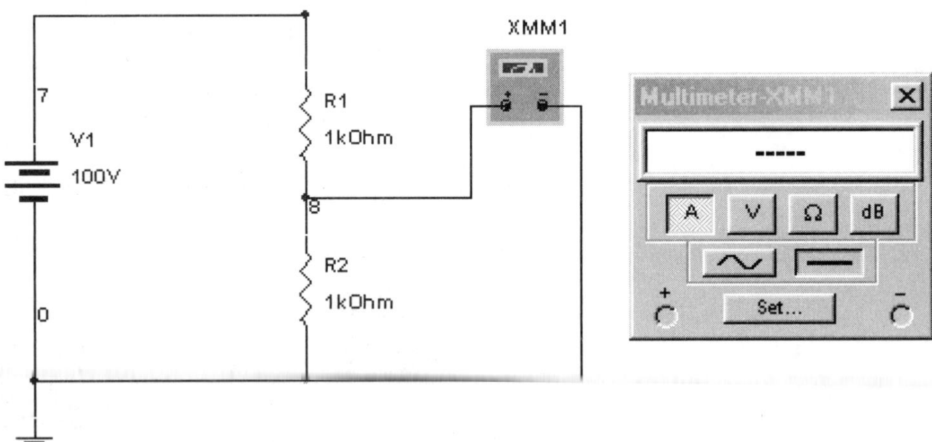


**Figure 9-3** Digital Multimeter Connection to Measure DC Voltage

5. Now take a look at the ohmmeter function of the digital multimeter.

   a. Open circuit file **09-01c1** (see Figure 9-4). In this circuit, the meter is connected to measure resistance. Notice the meter is measuring a resistor that is part of a circuit with the DC power source disconnected. It is necessary to disconnect the power source whenever a resistance measurement is being made. This has to be done in EWB, which is simulating real-world activity, as well as in the real world where protection of the meter is of prime importance. Activate the circuit. What is the measured resistance reading?

The measured resistance is __________ Ω.

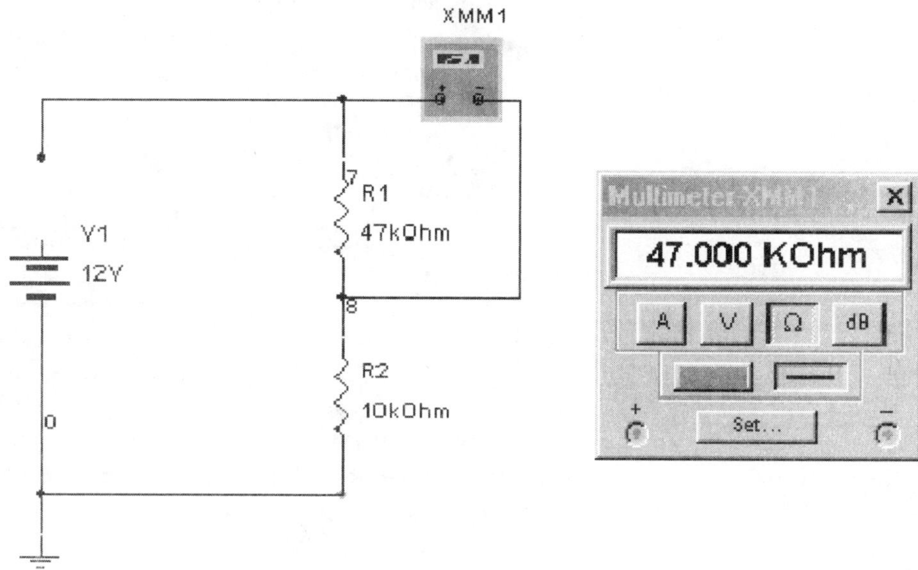


**Figure 9-4** Digital Multimeter Connection to Measure In-Circuit Resistance

b. Open circuit file **09-01c2** (see Figure 9-5). In this circuit, the meter is connected to measure resistance "out-of-circuit." Activate the circuit and determine the resistance of the resistor. The measured resistance is __________ Ω. When using the *MultiSIM*® program, the reading

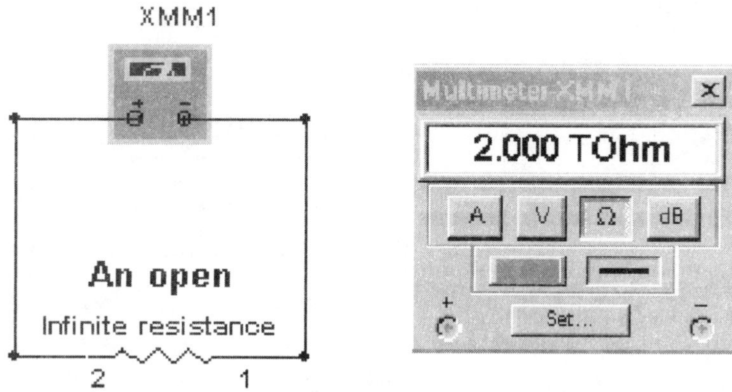


**Figure 9-5** Digital Multimeter Connection to Measure Infinite Resistance (an open circuit)

for infinite resistance will be **2Tohm** (Terra ohms = $1 \times 10^{+12}$). When using EWB Version 5, the indication for infinite resistance is the "**i**" on the display of the DMM. Electronics technicians usually refer to a measurement with infinite resistance as being an "**open**" circuit.

c. Open circuit file **09-01c3** (see Figure 9-6). The meter is connected to measure resistance. Again, notice the meter is measuring a resistor "out of circuit." Activate the circuit. What is the resistance reading?

The measured resistance is __________ Ω. Electronic technicians usually refer to a zero ohm measurement as being a "**short**" when the measurement is a fault condition.

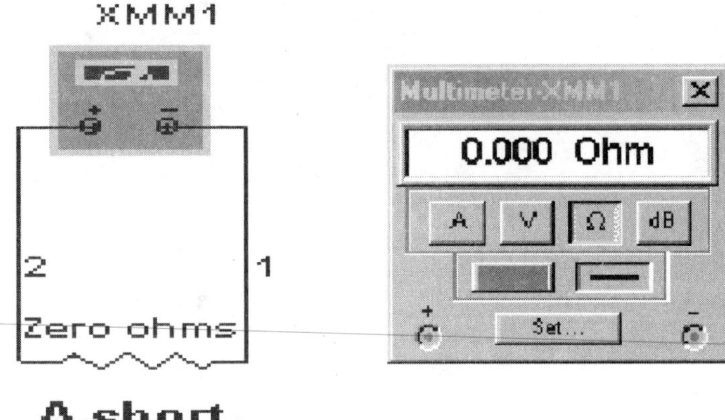


**Figure 9-6** Digital Multimeter Connection to Measure Zero Ohms (a short)

## Activity 9.2: Using the Digital Panel Meters in DC Circuits

1. The EWB digital panel meters are located in the **Indicators** menu on the taskbar. There are two types, the ammeter and the voltmeter. These can be used for DC or AC. Now the DC function will be discussed. Figure 9-7 presents the two types of meters used in the EWB program.

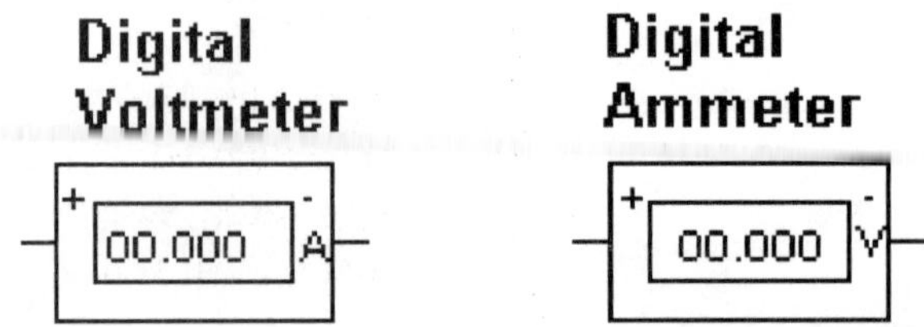


**Figure 9-7** Digital Panel Meters Found in *MultiSIM®*

2. The digital ammeter works essentially the same as the ammeter function on the DMM that has previously been used. One advantage to this type of meter, at least in EWB, is that as many of these panel meters as desired

can be brought out onto the workspace. With the multimeter, only one can be used at a time. The disadvantage with this type of meter is that it has only one function, to measure current. It does not measure resistance or voltage.

3. Open circuit file **09-02a**. This circuit is ready for installation of an ammeter between TPA and TPB. Go to the **Indicators** menu, open it, click on the ammeter symbol, wait for the menu to appear, click on **AMMETER_V, OK**, and then place the meter on the workspace. Notice that the positive sign is at the top of the ammeter representing the positive connection point for the meter. Now install the ammeter in the circuit between TPA and TPB. Activate the circuit and note the meter reading.

   The measured DC current reading is __________ mA.

4. Open circuit file **09-02b**. This circuit is ready for a voltmeter to be connected across $R_2$ by attaching the meter leads to TPA and TPB. Go to the **Indicators** menu, open it, click on the **VOLTMETER_V, OK**, and then place the voltmeter on the workspace. Notice that the positive sign is at the top of the voltmeter representing the positive connection for the meter. Now connect the voltmeter to TPA and TPB in the circuit. Activate the circuit and note the meter reading.

   The measured DC voltage reading is __________ V.

## Activity 9.3: Internal Resistance of Current and Voltage Meters

1. Meters frequently offer a load problem for electronic circuits, as previously discussed in Chapter 6, Activity 6.5. In a high-resistance circuit, a low-resistance voltmeter can adversely affect a circuit when placed in parallel with circuit components by providing an additional parallel path for current flow. A high-resistance ammeter can also adversely affect a circuit under test by placing additional and undesired series resistance into a low-resistance circuit. Either way, internal meter resistance always has to be taken into account whenever electronics circuits are being tested.

2. Open circuit file **09-03a**. In this circuit, the two types of voltmeters used in EWB are being used to measure the voltage drop across one of the resistors in the circuit. Activate the circuit and note the voltage reading on the connected DMM. What is it?

   The reading on the DMM is __________ V. Now, connect the panel meter

   M1 to TPA. The reading on the DMM has changed to __________ V. There is an obvious problem when the digital panel meter is added to the circuit. What is wrong? What can be done about it?

3. Move the mouse pointer to the DMM and right-click on the **Settings** block. Notice that the voltmeter resistance setting is set to 1 GΩ. Change the setting to 1 MΩ and reactivate the circuit. Now both meter readings are identical and "wrong." Obviously, the internal resistance setting is important in high-resistance circuits. Reset the multimeter internal resistance to 1 GΩ. Go to the digital panel meter. Now, double left-click on the meter, click on the **Value** tab, and notice that the meter resistance setting is set for 1 MΩ. This tells us that the meter, with an internal resistance of 1 MΩ, is placed in parallel in the circuit with a circuit resistor of 1 MΩ. There are parallel paths for current flow through the meters. The resistance paths formed by the digital panel meter and resistor are of equal value with equal currents flowing through each parallel path. Now, change the digital panel meter to 999 MΩ. Reactivate the circuit and the meter readings are equal and correct. Both of the meters have such high resistance that the small amount of current flowing through them can be ignored. The bulk of the current is flowing through the resistor.

4. Open circuit file **09-03b**. In this circuit, the two types of ammeters found in EWB are being used, alternately, to measure the current flow through a 0.5-Ω resistor. The switch is used to alternate the meter activity. Activate the circuit and determine the current reading through the DMM. The circuit current through the DMM is __________ A. Toggle the switch and determine the current through the digital panel meter. The current reading through the digital panel meter is __________ A. There is an obvious discrepancy between the two meters. What is wrong? What can be done about it?

5. Focus, first, on the digital panel meter. Open up the **Value** tab for the digital panel meter and change the internal resistance from 100 mΩ to 1 nΩ. Reactivate the circuit and check the two readings. They are now equal. The digital panel meter internal resistance in this circuit was too high and dropped excessive voltage.

## Activity 9.4: Calculating the Value of Shunt Resistors for Ammeters

1. The movement of an analog ammeter is rated in "the amount of current that will produce full-scale deflection of the needle," a visible left-to-right movement that indicates, in a linear fashion (generally), how much current is flowing through the meter. This current flow cannot be more than the amount necessary to produce full-scale deflection, or potentially, damage to the meter movement could result. The excess current, when it is necessary to measure more than rated full-scale current, has to be bypassed around the meter through a "**shunt**" resistor. The basic circuit for an ammeter shunt is shown in Figure 9-8. For example, if a specific ammeter has an internal resistance of 100 Ω for a full-scale needle deflection current of 1 mA, and 100 mA is being measured, then it is necessary to make sure that 99 mA bypasses the meter coil (this prevents damage).

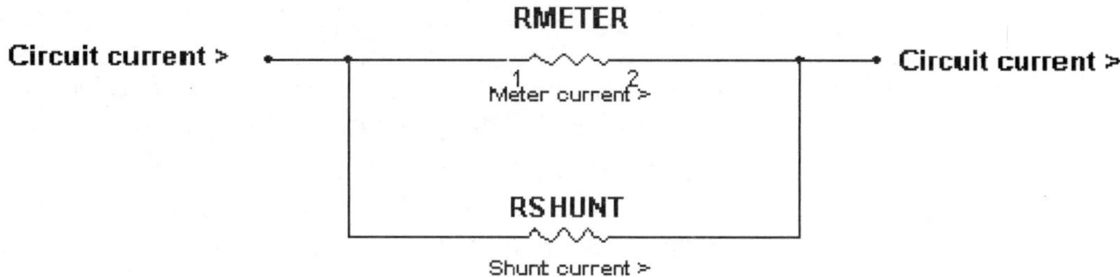


**Figure 9-8** An Ammeter Shunt Resistor Circuit

2. Open circuit file **09-04**. This circuit has a meter movement with a full-scale deflection rating of 1 mA. The meter movement has a resistance of 50 Ω. It should measure 100 mA for full-scale deflection. What value of shunt resistor would bypass the extra 99 mA of current? Use the formula, $R_{SHUNT} = I_{METER} \times R_{METER}/I_{SHUNT}$.

   The value of the shunt resistor is __________ Ω.

3. Change the value of the shunt resistor, $R_{SHUNT}$, to the calculated value. Activate the circuit and verify that the calculations are correct. $M_1$, representing the current through the analog meter movement should have 1 mA flowing through it, and $M_2$, representing the current through the meter shunt resistor, should have 99 mA of current flowing through it.

## Activity 9.5: Calculating the Value of Multiplier Resistors for Voltmeters

1. The movement of an analog voltmeter is rated in the same manner as an ammeter because the basic meter movement reacts to current flow through it. The difference between the ammeter circuit and the voltmeter circuit is that instead of bypassing potentially damaging current around the meter as with ammeters, "**multiplier**" resistors are placed in series with the voltmeter meter movement to drop excess voltage and to keep the current in the proper range. The basic circuit for a voltmeter is shown in Figure 9-9. For example, if a specific voltmeter has an internal resistance of 100 Ω for a full-scale needle deflection current of 1 mA, the meter would drop 100 mV. When measuring 10 V and it is necessary to obtain full-scale deflection at that voltage, then the multiplier resistor has to drop 9.9 V.

2. Open circuit file **09-05**. This circuit has a meter movement rated at 1 mA for full-scale deflection and measures 50 Ω of resistance. The voltage under test is 100 V and that should cause full-scale deflection. What value of multiplier resistor would drop the extra voltage? Use the formula, $R_{MULTIPLIER} = (V_{RANGE}/I_{FULL\text{-}SCALE}) - R_{METER}$.

   The calculated value of the multiplier resistor is __________ Ω.

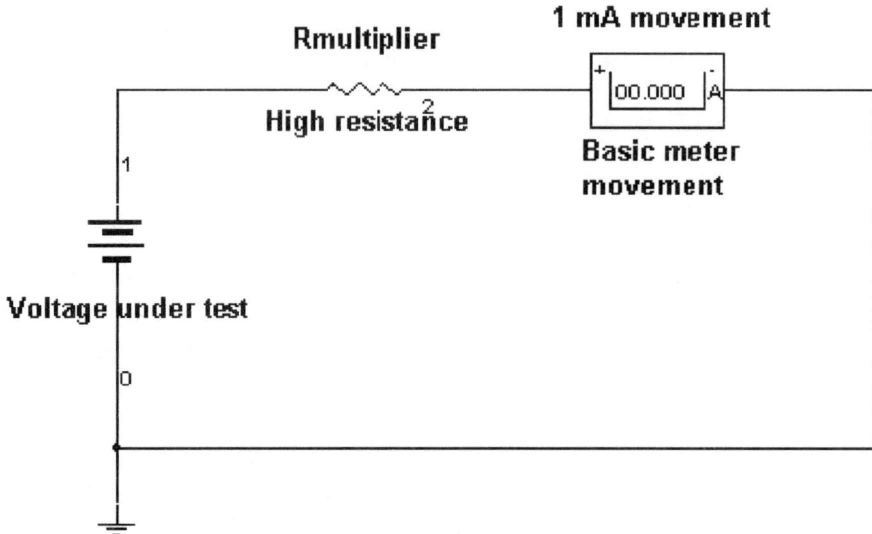


**Figure 9-9** A Voltmeter Multiplier Resistor Circuit

3. Change the value of the multiplier resistor, $R_{MULTIPLIER}$, to the calculated value. Activate the circuit and verify that the calculations are correct. M1, representing the voltage being measured should have 1 mA flowing through it with its rated voltage drop across. The rest of the voltage should be dropped across the multiplier resistor.

## Activity 9.6: The Oscilloscope as a DC Test Instrument

1. The **Oscilloscope** (scope or o-scope) is one of the most widely used and capable test instruments available to the electronics technician. It can be used to measure DC as well AC, but is most commonly used for its AC measurement capabilities. However, it will measure DC very well. When you first look at an oscilloscope you might be overwhelmed by all of the switches and controls (see Figure 9-10). The scope in EWB may not be as complex to look at as the scopes in the lab or on the test bench, but it will accomplish all that is necessary with EWB circuits and problems.

2. Open circuit file **09-06a**. Notice that the scope is divided into four sections; these sections are the functional areas of the scope. The four sections are the **Timebase**, **Channel A**, **Channel B**, and the **Trigger** section. Each has a specific function to perform so the oscilloscope is able to do its task—to display the operating characteristics of various types of electronics circuits. In this circuit, the scope is connected to a DC power source and is prepared to display a DC voltage.

3. Activate the circuit and observe the line that crosses the scope from left to right. Notice the X-axis line that crosses in the center of the scope. This line is calibrated to provide an indication of time relationships in the

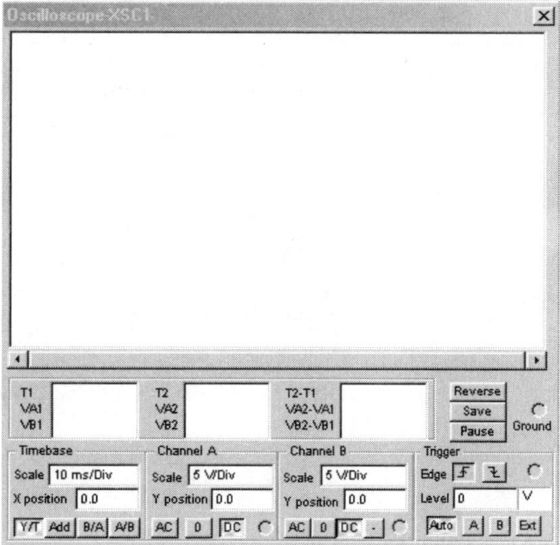


**Figure 9-10** The Faceplate of the EWB Oscilloscope

X-axis. The Channel A section, which is connected to the DC power source, is set for **5 V/Div**, which means that a large division on the vertical axis is equal to 5 volts. The line crosses the scope on the first division above the center, which means that the voltage being measured is equal to $1 \times 5\ V = 5\ V$. The X-axis crosses the screen in the center of the display normally representing 0 V.

4. Change the voltage of the DC power source to 10 V. Activate the circuit and observe where the DC line crosses the vertical axis. The line crosses

   the scope on the __________ large division line above 0 V which is equal to

   __________ V.

5. Between the large divisions are five smaller divisions on the horizontal axis and spaces on the vertical axis. Each of these is equal to 20% of the value of one division. If one division were equal to 5 V on the vertical axis, then each smaller division would be equal to 1 V. Change the voltage output of the DC power source to 8 V. Activate the circuit and observe that the DC line crosses the vertical on the Y-axis about eight divisions above the center line.

6. Change the output of the DC power source to 100 V. The Channel A input setting has to be changed from 5 V/Div to 50 V/Div. Activate the circuit and observe where the DC line crosses the Y-axis again.

   The line crosses the Y-axis on the __________ division line which is equal

   to __________ V.

## ● *Troubleshooting Problem:*

7. Open circuit file **09-06b**. The scope is already connected to the circuit under test. What is the value of the voltage at TPA?

   The voltage at the TPA is __________ V (you will have to change the Channel A V/Div setting). If the horizontal line is off the top of the scope, then the voltage is more than can be displayed by the Channel A V/Div setting. Increase the setting until the DC line (trace) is displayed. In this

   case, the proper V/Div setting is __________ V/Div.

# 10. Electromagnetic Devices

> ***References***
> *Electronics Workbench®, MultiSIM* Version 6
> *Electronics Workbench®, MultiSIM* Version 6 Study Guide

**Objectives** After completing this chapter, you should be able to:

- Use simple and variable inductor libraries in *Electronic Workbench®*.
- Use standard and nonlinear transformer libraries.
- Use relays to remotely control circuits.
- Connect DC motors and vary their output speed.

## Introduction

Electromagnetic devices are used throughout the electronics industry in component applications ranging from simple devices such as transformers, relays, and motors, to complex systems that depend on electromagnetic principles to attain desired effects. The principles of magnetism, electromagnetism, electromagnetic radiation, induction, and other related phenomenon have, over a period of several hundred years, infiltrated all aspects of modern industry and society.

The development of electrical power through the use of electromagnetic generators; the distribution of that electrical power through the capabilities of transformers; and the industrial application of that power, particularly with DC and AC motors, has brought about today's modern society. A return to the pre-electromagnetic days is almost impossible today. Third world countries around the globe are in a catch-up mode as they attempt to achieve industrial and economic independence through modern industry and develop the means by which that industry primarily operates today—the electric motor and AC electrical power.

In this section, the application of various electromagnetic devices found in the EWB software will be discussed. These devices represent only a small selection of the many components and systems available today. The technician needs to understand the theory as well as the applications of these magnetic, inductive, and electromagnetic devices. You will find at least some of these various applications in practically every piece of electronics equipment that you work on. Figure 10-1 displays the various electromagnetic devices found in EWB.

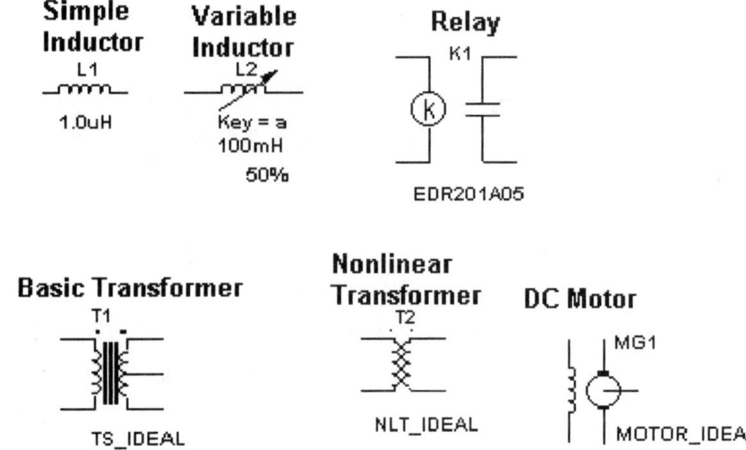


**Figure 10-1** Electromagnetic Components Found in EWB

## Activity 10.1: Using the Inductors Found in EWB

1. There are two types of **inductors** (coils) in EWB *MultiSIM®*, the simple inductor and the adjustable inductor. They are both found in the **Basic** menu. At this point, it is necessary to study inductors before attempting their application in electronics circuits. They are primarily used in AC circuits and do not react in DC circuits. Generally, the only characteristic to be concerned about regarding inductive components in DC circuits is their DC resistance, consisting primarily of the resistance of the wire that makes up the device.

2. Open circuit file **10-01**. These are the three types of simple inductors (coils) found in EWB (see Figure 10-2). Left-click on the simple inductor first and then click on **Replace** for the menu of inductors. Notice that there is not much you can do with this device except change its inductance value. Next, if you left-click on the adjustable inductor you will observe that it is similar to the simple inductor. Adjustment of the inductance of the variable inductor can be accomplished in a manner similar to that used with potentiometers, by using a selected key on the computer keyboard and changing the value of inductance.

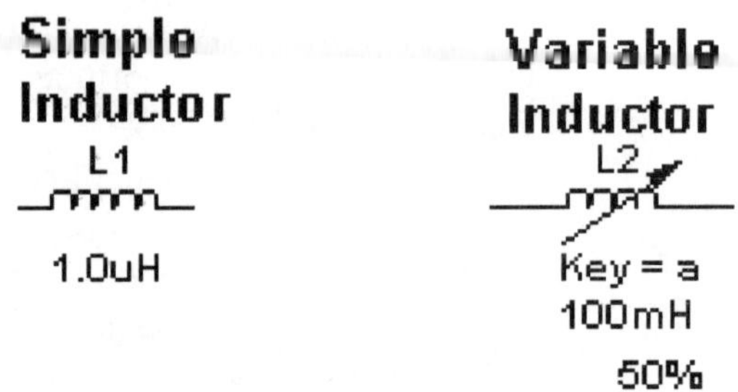


**Figure 10-2** Inductors Used in *Electronics Workbench®*

## Activity 10.2: Using EWB Transformers

1. There are two types of transformers used in EWB *MultiSIM*®, the standard transformer and the nonlinear transformer. Both are found in the **Basic** menu. Transformers are primarily used in AC circuits and do not normally operate in DC circuits. The only characteristic of concern in DC applications is the DC resistance of the transformer consisting of the resistance of the wire that makes up windings of the device. The transformers presented in this circuit are more complex than the EWB components that have been studied up to this point. One big difference is that when you left-click on the transformer symbol from the Basic library, a "browser" presents a menu of transformer choices as shown in Figure 10-3. The transformer usually used is the "ideal" transformer. The other transformers in the library have electrical characteristics that make them different from the ideal transformer. These characteristics will be studied later. There is also an "**edit**" selection that allows for change of the electrical characteristics of the transformers when necessary. Take care when leaving the program; many times the program will question whether you wish to modify the component library after you have edited a circuit containing a transformer. The answer is **always "no."** The existing defaults need to stay the same as they are to avoid confusion later when it is not possible to remember what the defaults were.

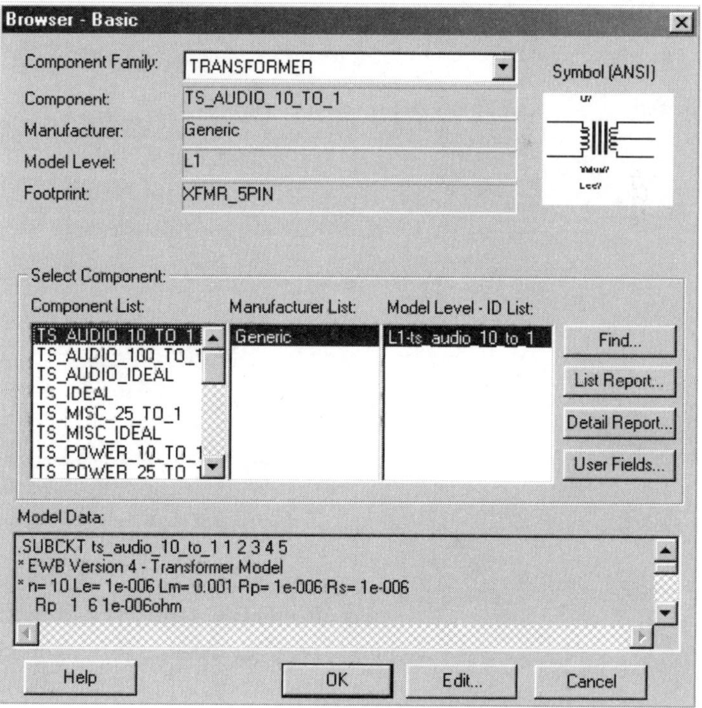


**Figure 10-3** Transformer Library in *Electronics Workbench*®

2. Open circuit file **10-02**. In this circuit, a transformer is installed, using a menu choice with a ten-to-one step-down ratio. Use the DMM to measure the voltage of the primary and secondary.

   Measured $V_{PRIM}$ = __________ V and $V_{SEC}$ = __________ V.
   Did the voltage readings verify the step-down ratio?

## Activity 10.3: Using the EWB Relay Component

1. Relays are electromagnetic switching devices with coils that are energized by current flow through the coil. The energizing action of the current flow causes a mechanical contact transfer to take place. There is only one of these components in the **Basic** library (see Figure 10-4).

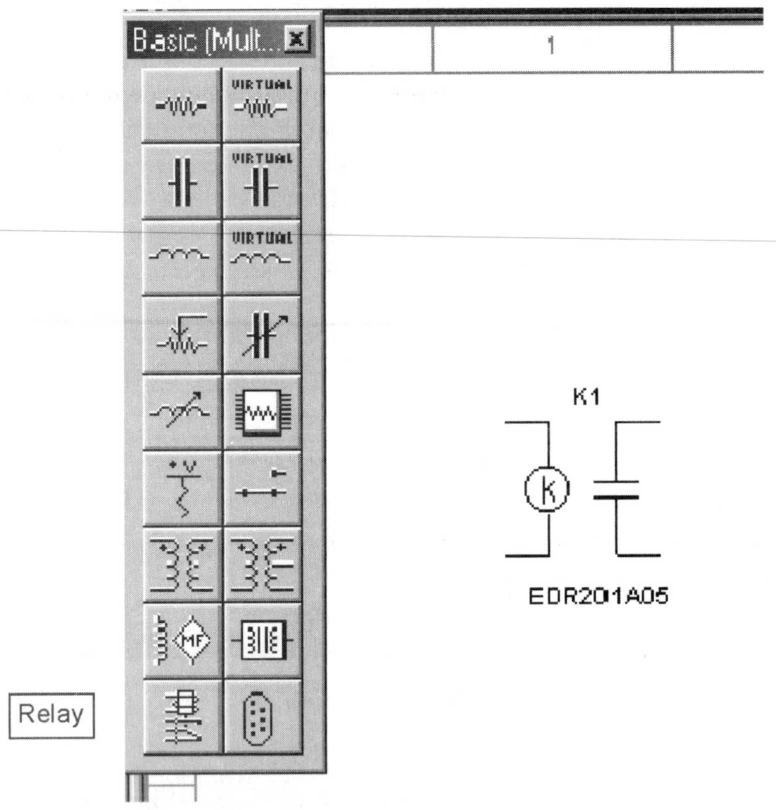


**Figure 10-4** Basic Menu Showing the Relay

2. Open circuit file **10-03**. This circuit is designed to operate a relay and control another circuit. When the circuit is activated and S1 is closed, the relay switches on; its contacts close the controlled circuit containing the lamp. As a result of the switching action, the lamp, which is electrically isolated from the control circuit that energizes the relay, is turned on. Notice the switching action of the relay contact when the switch is actuated. This is one of the primary uses of a relay, to isolate one power source

from another: for example, isolating a DC power source from an AC power source. Install the ammeter and measure the current flowing through the lamp. $I_{Lamp}$ = __________ A.

## Activity 10.4: Using the EWB DC Motor

1. Motors are rotating electromagnetic devices that change electrical energy into mechanical energy. In EWB there is one motor, a DC motor, that does not do anything mechanical. It just acts as a load and provides a "revolutions per minute (RPM)" signal output to a panel meter to indicate its speed in relation to the input voltage/current. This is an excellent tool to develop circuits to control DC motors and similar loads where output RPM is important.

2. Open circuit file **10-04**. This circuit demonstrates the operation of the EWB DC motor. When the motor is energized, the voltmeter displays a reading relative to the speed of the motor. Varying armature current changes the speed of the DC motor. Speed variation and control is accomplished in this circuit by the potentiometer.

3. Vary the potentiometer according to the parameters of Table 10-1 and watch the voltmeter reading change; the reading on the voltmeter can be directly correlated with the RPM of the motor (not a real voltage output). Record the RPM data in Table 10-1.

| | **DC Motor Speed Correlated With Potentiometer Setting** | | | | | | |
|---|---|---|---|---|---|---|---|
| Potentiometer % Setting | 0% | 10% | 25% | 50% | 75% | 90% | 100% |
| DC Motor RPM | | | | | | | |

**Table 10-1** DC Motor Speed Control

4. At what (percentage) setting of the potentiometer is it possible to achieve maximum RPM out of the DC motor?

   Maximum RPM is achieved at __________% potentiometer setting.

5. If it is desired to slow the motor down more than the minimum speed possible with the 1-kΩ potentiometer, how can it be accomplished? How can the motor be slowed down more than is possible with the present circuit? The motor could be slowed down even more than its present speed by

   ______________________________________________________.

# 11. Alternating Voltage and Current

**_References_**

*Electronics Workbench®, MultiSIM* Version 6

*Electronics Workbench®, MultiSIM* Version 6 Study Guide

**Objectives** After completing this chapter, you should be able to:

- Use an oscilloscope to obtain voltage, frequency, and period (time) measurements of an alternating current (AC) waveform.
- Explain the relationships between frequency and period of an AC waveform.
- Discuss relationships between peak, peak-to-peak, RMS, average, and instantaneous values of AC waveforms.
- Understand phase relationships between waveforms.
- Harmonize relationships between AC waveforms.
- Determine characteristics of EWB measuring instruments in AC circuits.
- Analyze voltage, current, resistance, and power in resistive AC circuits.

## Introduction

Alternating current is used throughout the electrical and electronics industry in applications involving transformers, relays, and motors and also in complex systems that depend on electromagnetic principles to attain desired effects. The entire industrial base of modern society is based on the generation, distribution, and use of alternating current.

Direct current is unidirectional, generally fixed in value, and has polarity while AC varies in direction, value (amplitude), phase, and polarity. A key advantage of AC over DC is the ease in which voltage levels can be stepped-up or stepped-down by means of transformers—an easy change of voltage amplitude, which greatly benefits power distribution.

In EWB, either the function generator from the **Instruments** menu or the AC signal source from the **Sources** menu can provide AC signals to the circuits (see Figure 11-1). The voltage and current outputs of signal sources are peak outputs rather than RMS outputs. The meter readings will be RMS readings and the oscilloscope will display peak-to-peak voltages. The function generator output is a peak output when one of the output leads is connected to the common (middle) terminal. If the output is connected between the positive and the negative terminals, the output will be peak-to-peak (two times peak).

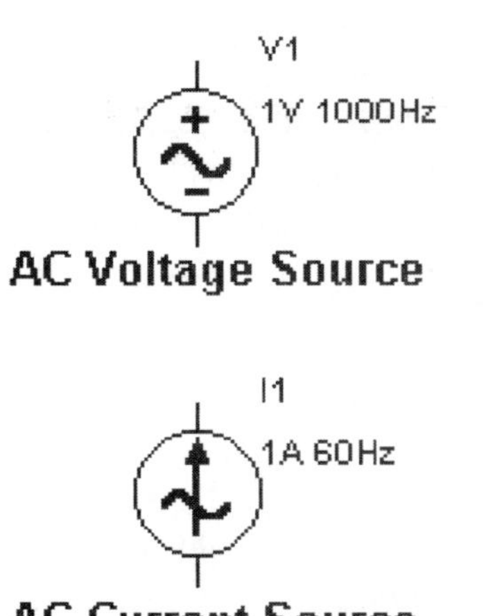


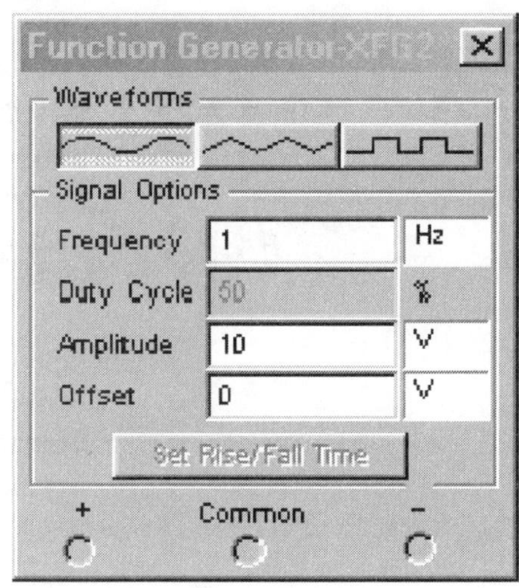


**Figure 11-1** AC Signal Sources Found in EWB

## Activity 11.1: AC Waveforms

1. The function generator produces three types of signal waveforms: sinusoidal (the sine wave), triangular, and square.

2. Open circuit file **11-01**. In this circuit an oscilloscope is connected to the output of the function generator. We are using the oscilloscope to observe the three types of waveforms produced by the function generator. Later, other scope functions will be studied. Enlarge the scope and the function generator. Activate the circuit and use the pause button near the on-off switch or the pause button on the oscilloscope to stop the waveform. Left-click on the waveform task bar at the top of the function generator, one at a time, and observe the three types of displayed waveforms. The three types of waveforms are ______________, ______________, and ______________. To stop the scope trace from "free-running," click the pause button on the oscilloscope panel before activating the circuit.

## Activity 11.2: Measuring Time and Frequency of AC Waveforms

1. Previously we learned about the display of DC voltages on an oscilloscope. That investigation primarily consisted of learning about Y-axis displacement (vertical displacement) as a means of measuring DC voltage levels. In this chapter, interpretation of the various types of AC waveforms displayed on the oscilloscope and an understanding of the meaning of these various types of scope data will be developed.

2. The time relationship of the AC waveform is displayed on the X-axis (horizontal displacement). In this case the **Timebase** section of the scope is the area that is under consideration. The time base section of the scope

is adjusted to determine various aspects of the AC waveform in relationship to time. Observe that the time base is set for 200 μs/div, which means that one large division of horizontal deflection on the X-axis is equal to 200 microseconds (μs) in time. If the displayed waveform took seven divisions to complete a cycle of operation, then the time required to complete that cycle of the waveform would be 200 microseconds × 7 divisions = 1.4 ms.

3. Open circuit file **11-02a** and activate the circuit. Notice that the sine wave displayed on the oscilloscope takes 5 divisions to complete one cycle of operation. For this type of waveform, one cycle of operation is from the beginning of the waveform on the left side of the scope face (where the trace goes up) to the end of one cycle of operation where the waveform starts to go up again. The waveform rises at the left, stays high (flat top) for 2.5 divisions, falls to a minimum point (below the X-axis), stays there for 2.5 divisions and starts the cycle over again by rising once again. Notice that, in Figure 11-2, the waveform rises two divisions above the X-axis, which represents a voltage level of 10 V/Div × 2 divisions = +20V. Then the waveform falls to a point that is two divisions below the X-axis (which represents zero volts) indicating a voltage level of –10 V/Div × 2 divisions = –20 V. The waveform has a top-to-bottom (peak-to-peak) value of 40 V, from –20 V up to +20 V.

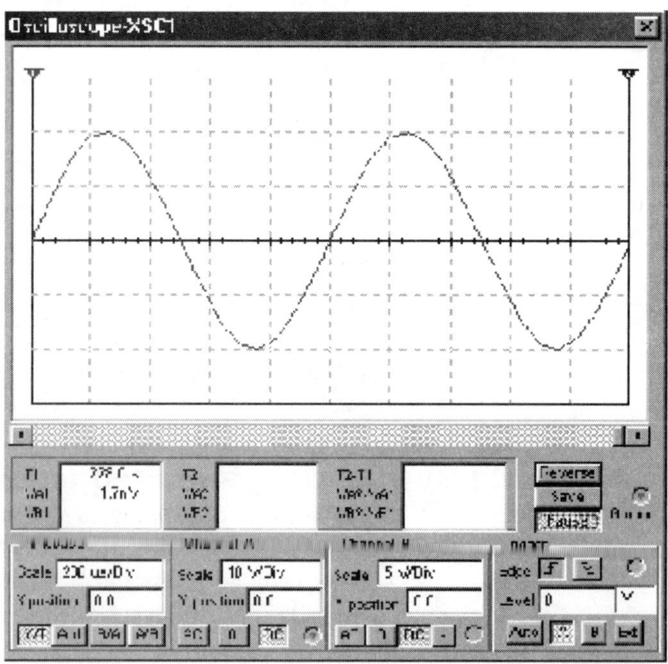


**Figure 11-2** The Expanded Oscilloscope View

4. The formula **f = 1/t** develops the relationship between the time for one cycle of the displayed signal and how often a repetitive waveform occurs per second (frequency). The unit of measurement for **frequency** is the **hertz**. If a

waveform occurs ten times per second, it occurs at a frequency of 10 Hz. In this circuit, the waveform under observation takes 5 × 200 μs = 1 ms in time for one cycle and, thus, its frequency is 1/0.001 or 1000 Hz (1 kHz).

5. Open circuit file **11-02b** and activate the circuit. Enlarge the scope and observe the 1-kHz signal. Notice the red and blue vertical cursors positioned at the beginning and the end of the second sine wave. Try moving them with the mouse. There are three windows under the scope display (see Figure 11-3). Window 1 on the left displays Cursor 1 information, T1—time into the sweep (1 ms from the left of the scope face) and V1—the voltage at the point where the cursor is located (where it crosses the signal display). Window 2 displays Cursor 2 information, which is the same kind of information as Cursor 1. Window 3 displays time (T2 – T1) and voltage (VA2 – VA1) differences between the two cursors. These cursors are valuable because they can be positioned on Channel A signals and measure time differences between two points as well as measuring voltage differences.

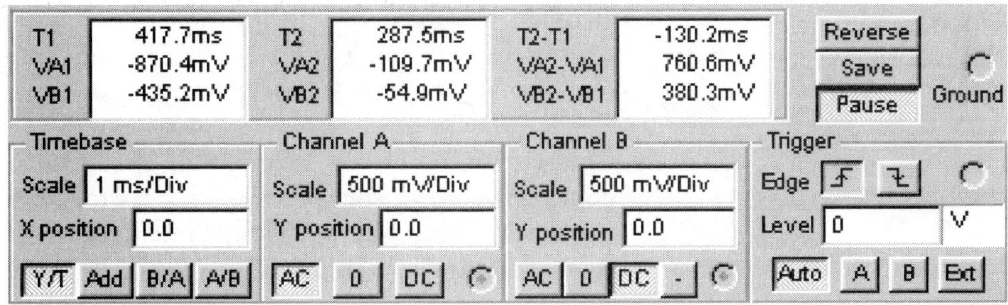


**Figure 11-3** The Windows of the Expanded Oscilloscope

6. In the case of this signal, the cursors are positioned at the beginning and the end of a complete sine wave and Window 3 is displaying the time difference. The time difference T2 – T1 is __________ ms and the frequency of the signal is __________ kHz (f = 1/t).

## ● *Troubleshooting Problems:*

7. Open circuit file **11-02c**. Activate the circuit and determine the frequency of the square wave displayed on the oscilloscope. The time for one cycle is __________ ms. The frequency of the square wave is __________ Hz. Notice that this is a DC signal; it never goes below the 0 V reference line in the middle of the Y-axis.

8. Open circuit file **11-02d**. Using this circuit, determine the frequency of the sine wave displayed on the oscilloscope. The time for one cycle is __________ ms. The frequency of the sine wave is __________ Hz.

9. Open circuit file **11-02e**. In this circuit, adjust the time base setting to observe the period of one sine wave. Then determine the frequency of the sine wave displayed on the oscilloscope. The time for one cycle is __________ ms. The frequency of the sine wave is __________ Hz.

## Activity 11.3: Measuring RMS, Average, Peak, Peak-to-Peak, and Instantaneous Values of AC Waveforms

1. It is easy to determine the voltage level of DC voltages on an oscilloscope because they are displayed as a straight line crossing the Y-axis at a determinable point. AC is harder to deal with because it is a continually changing waveform and is very difficult to pin down to a specific value of voltage. There are other points of concern for the technician in the area of instrumentation: What are the voltage values? Are they accurate and are all of the instruments saying the same thing? In this activity, you will gain a better understanding of RMS, average, peak, peak-to-peak, and instantaneous values of waveforms.

2. Open circuit file **11-03a**. Activate the circuit and observe the straight line going across the display. At the bottom of the Channel A section notice that the channel is set for "**0**" rather than AC or DC. When the channel is set for "0," the "zero" reference for the screen is displayed in the center of the screen on the X-axis (this can be adjusted by the offset arrows, moving the trace up and down as necessary). Click on the "**DC**" setting and observe the waveform. This waveform starts at __________ V and rises to __________ V. Its frequency is __________ kHz. Again, this is a DC waveform because it never goes below the zero reference line.

3. Open circuit file **11-03b**. Activate the circuit and expand the scope. Notice that Cursor 2 and Cursor 1 are positioned at the highest and lowest points on the square wave signal. Window 2 indicates that the waveform rises to about __________ V and Window 1 indicates that the waveform falls to about __________ V.

4. The portion of the waveform above the zero reference point is called the "**positive peak**" voltage and the waveform below the zero reference would be called the "**negative peak**" voltage. When referring to the overall waveform, the term **peak-to-peak** is used. In this circuit, the peak positive voltage = +__________ V, the peak negative voltage = –__________ V, and the peak-to-peak voltage = __________ V.

5. Open circuit file **11-03c**. Activate the circuit and determine the peak and peak-to-peak values of the displayed signal.

   In this circuit, the peak voltage = +__________ V or –__________ V and

   the peak-to-peak voltage = __________ V.

6. **Instantaneous** voltage is the voltage at any specific point on the AC signal at that instant in time. When using the cursors in the expanded display, the $V_1$ and $V_2$ values in Windows 1 and 2 represent the instantaneous voltage value of the displayed signal wherever the cursor is located at that point on the Channel A waveform.

7. Open circuit file **11-03d**. Activate the circuit, expand the scope, and click on the pause button to stop the sweep. Place the red cursor on the positive peak of the waveform. Read the voltage of the positive peak (VA1) from the first window to the right directly below the display area of the scope.

   The value of the positive peak is $V_1$ = __________ V. Place the blue cursor on the negative peak of the waveform. Read the voltage of the negative peak (VA2) from the second window.

   The value of the negative peak is $V_2$ = __________ V. To determine instananeous values at a particular angular point move either cursor to that point and readout the voltage.

8. When dealing with **RMS** voltage values, the EWB oscilloscope displays peak-to-peak signals. RMS values cannot be obtained from waveforms displayed on the oscilloscope without performing some mathematical calculations. RMS = 0.707 × peak voltage. So, a 100-V peak signal would equal 70.7 VRMS.

9. When dealing with **Average** voltage values, again it is necessary to go to mathematical calculations to determine the value.
   Average voltage = 0.637 × peak voltage. So a 100-V peak signal would equal 63.7 V average.

## Activity 11.4: Phase Relationships between Waveforms

1. A point on an AC signal is often referred to in degrees of angular rotation; for example, the 30° point of a signal. When discussing **phase difference** between sine waves and **phase shift** of a signal in an electronics circuit, the term used to discuss the amount of phase difference or phase shift is also in degrees of angular rotation. The oscilloscope is an excellent tool to determine phase differences; Channel A can observe one signal and Channel B can observe another signal and then compare their relationship in time. With signals, when you want to determine the amount of phase differences, the signals have to be of the same signal frequency, but can differ in voltage levels.

2. Open circuit file **11-04a**. In this circuit, there are two signal sources ($V_1$ and $V_2$). The wire from $V_1$ to Channel A on the oscilloscope is red and that signal will be red on the scope display. The wire from $V_2$ to Channel B on the oscilloscope is blue and that signal will be blue on the scope display. These colors will be standard operating procedure for the rest of this study: Channel A input will be red and Channel B input will be blue. Activate the circuit

   and observe that the two signals appear to be __________° apart in phase relationship.

3. Each value of phase shift, or phase difference, with most oscilloscopes is always an approximation. You determine, as closely as possible, the value in degrees of angular rotation for each division on the X-axis, and approximate the phase difference from that knowledge. In this circuit, one sine wave is ten divisions in length resulting in one division being equal to 36° in angular rotation. If there were four divisions between the two signals from the same relative point on the signal, then the amount of

   phase difference would be __________°.

4. Open circuit file **11-04b**. Once again, the value of one division on the X-axis is 36°. In this circuit the phase difference is less than 180°. What is the

   approximate phase difference? The phase difference is __________°.

## Activity 11.5: Harmonic Relationships between AC Signals

1. A harmonic signal is a signal that is a multiple of a reference signal. For instance, the second harmonic of a 1-kHz signal would be 2 kHz. A third harmonic would be 3 kHz.

2. Open circuit file **11-05a**. Activate the circuit and observe that there are two blue sine waves for every red sine wave. Thus, the Channel B signal

   is a __________ harmonic of the Channel A signal.

3. Open circuit file **11-05b**. Activate the circuit and observe the signal differences between the two oscilloscope input channels. The Channel B

   signal is a __________ harmonic of the Channel A signal.

## Activity 11.6: Characteristics of EWB Measuring Instruments in AC Circuits

1. The voltage, current, and power values in AC circuits are usually dealt with in terms of RMS values. Be careful to make sure that when measurements are being made of values other than RMS values, there is not a

mixture of values, such as mixing average, peak, and RMS voltage values. Any mixture of measurement values will cause an error in the calculations. The oscilloscope display has to be approached with caution because the displayed data is in the form of peak or peak-to-peak values and most measuring instruments other than the oscilloscope are RMS reading instruments. When a signal is being measured with a voltmeter, the value is usually an RMS value. When measuring the same signal with an oscilloscope the observed signal is a peak or peak-to-peak value.

2. Most pieces of measurement equipment are RMS reading instruments. The technician has to know what the various signal source output voltages are in terms of RMS, peak-to-peak, peak, and average. The determination of what unit of measurement the various EWB instruments refers to is an important step.

3. Open circuit file **11-06a**. The help menu for the multimeter states that the DMM is an RMS reading instrument. It is connected to an AC signal source from the **Sources** menu. Activate the circuit and notice that the output voltage of the AC signal source is not the same as the voltage reading on the DMM. Why is the DMM reading different from the signal source stated output? The reason that the DMM reading is different is

   because ______________________________________________.
   What would you expect the peak-to-peak voltage to be on an oscilloscope? The oscilloscope would display a peak-to-peak signal of about

   __________ V.

4. The final instruments to learn are the voltmeters and ammeters in the **Indicators** menu. According to the help function, these meters are RMS reading devices. Open circuit file **11-06b** and activate the circuit. Note the meter readings. Connect $M_1$ to TPA and TPB. Reactivate the circuit and observe that obviously there is something wrong with meter $M_1$. To solve the problem, right-click on $M_1$, left-click on **Component properties**, and change the meter **Mode** from **DC** to **AC**. Reactivate the circuit

   again and all is well; $M_1$ reads __________ $V_{RMS}$ as it should. Whenever it is necessary to use these meters to measure AC, make sure that the meter mode is changed from DC to AC.

## ● *Troubleshooting Problem:*

5. Open circuit file **11-06c**. Activate the circuit. There is a problem in the circuit; the current through $M_2$ is wrong according to circuit calculations. Determine the problem and correct it.

   The problem is ______________________________________________

   ______________________________________________________________.

## Activity 11.7: Voltage, Current, Resistance, and Power in Resistive AC Circuits

1. The parameters of voltage, current, resistance, and power in AC resistive circuits are similar to the same parameters found in DC circuits regarding the relationships between Ohm's law, Kirchhoff's laws and the power formulas.

2. Open circuit file **11-07a**. Activate the circuit and determine the RMS value of the voltage drop across R1 with the attached voltmeter. Connect the oscilloscope to the circuit and verify the calculation.

   $V_{R1}$ = __________ $V_{RMS}$.

   What is the peak-to-peak value of the signal? V = __________ $V_{PP}$.

3. Open circuit file **11-07b**. Activate the circuit and determine the parameters of the circuit using the DMM, the oscilloscope, and your calculator. Record the data in Table 11-1.

| | Voltage | | | | Current | | | |
|---|---|---|---|---|---|---|---|---|
| | $V_P$ | $V_{PP}$ | $V_{RMS}$ | $V_{AVG}$ | $I_P$ | $I_{PP}$ | $I_{RMS}$ | $I_{AVG}$ |
| $R_1$ | | | | | | | | |
| $R_2$ | | | | | | | | |
| $R_3$ | | | | | | | | |
| Total | | | | | | | | |

**Table 11-1** Circuit Parameters of a Resistive AC Circuit

### • *Troubleshooting Problem:*

4. Open circuit file **11-07c**. Calculate circuit parameters and enter the data into Table 11-2. Then, activate the circuit and compare the calculations with the operating parameters of the circuit. There is a problem; what is it?

   The problem is ______________________________________________

   ______________________________________________.

   The circuit current measures about 12 mA. It should be __________ mA.

| | Voltage | | | | Current | | | |
|---|---|---|---|---|---|---|---|---|
| | $V_P$ | $V_{PP}$ | $V_{RMS}$ | $V_{AVG}$ | $I_P$ | $I_{PP}$ | $I_{RMS}$ | $I_{AVG}$ |
| $R_1$ | | | | | | | | |
| $R_2$ | | | | | | | | |
| $R_3$ | | | | | | | | |
| $R_4$ | | | | | | | | |
| $R_5$ | | | | | | | | |
| Total | | | | | | | | |

**Table 11-2** Circuit Parameters of Resistive AC Circuit

# 12. Inductance and Inductive Reactance

> ***References***
> *Electronics Workbench®, MultiSIM* Version 6
> *Electronics Workbench®, MultiSIM* Version 6 Study Guide

**Objectives** After completing this chapter, you should be able to:

- Use a test circuit to measure the inductance of an unknown inductor.
- Calculate and measure inductance in series and parallel inductive circuits.
- Calculate and measure inductive reactance and impedance in a circuit.
- Determine phase relationships of current and voltage in an inductive circuit.
- Calculate the Q and power losses of an inductor.
- Operate circuits containing inductance.
- Troubleshoot inductive circuits.

## Introduction

**Inductance (L)** is one of the basic phenomena of electronics that is used in all types of electronics equipment throughout the field of electronics. There are many components and electronic circuits using inductance and inductive characteristics. Inductance is defined as the property in a circuit that causes opposition to a change in current flow and has the ability to store a charge in an electromagnetic field. An **inductor** is a component that has inductance and is also called coils or chokes. The basis for an understanding of inductance is dependent upon an understanding of electromagnetic induction.

The definition of inductance—opposition to a change in current flow—tells us that inductance is an opposing element in alternating current circuits where there is a constant change of direction of current flow. In DC circuits, inductance has little effect except during the turn on and the turn off cycle or in the case of varying voltage levels such as pulsating DC or DC square waves.

Inductors have many uses in electronics. Typical uses for inductors as a component are in "tuned" circuits in radio and television circuits, as filters in electronic circuits to remove undesired signals and characteristics, in automotive applications such as the ignition coil, and many other usages. Inductance is found in transformers, motors, relays, and other electronic components.

Another characteristic of an inductor is its ability to store a charge. The charge is stored in the electromagnetic field that is built up around the component as a result of current flow. In a DC circuit, the electromagnetic field charge is built up around the inductor and remains there until power is removed. In an AC circuit, the electromagnetic field charge is constantly changing as the current continuously changes its direction of flow. Also, in an AC circuit, an inductor attempts to prevent the change in current flow while the energizing force, the voltage, is changing. As the voltage tries to increase or decrease and the current attempts to follow suit, the stored energy in the established electromagnetic field attempts to keep current from changing, thus opposing the change in current flow. The amount of opposition is dependent upon the strength of the charge stored in the electromagnetic field.

The amount of opposition to change in current flow that an inductive device offers to that attempted change in current flow is very important in an AC circuit. The term for the "amount of opposition" to a change in current flow is **Inductive Reactance**. Inductive reactance (**$X_L$**) is measured in ohms in a manner similar to resistance. For example, $X_L = 10\ \Omega$.

Wire has resistance and coils are usually constructed of many turns of wire. In EWB, the coils are referred to as "**ideal**" inductors and they have no resistance. The EWB inductors are found in the **Basic** library on the taskbar.

## Activity 12.1: Measuring Inductance with an Inductance Test Circuit

1. In Chapter 1, the various types of inductive devices were introduced. In this study, an inductive component, the coil, will be studied primarily under AC operating conditions. A coil is usually many turns of wire wrapped in a circular fashion around a coil form or core device.

2. Open circuit file **12-01a**. In this circuit a test circuit is used to measure the inductance of an unknown coil. This test circuit is designed using EWB components to test the unknown coil (see Figure 12-1). The

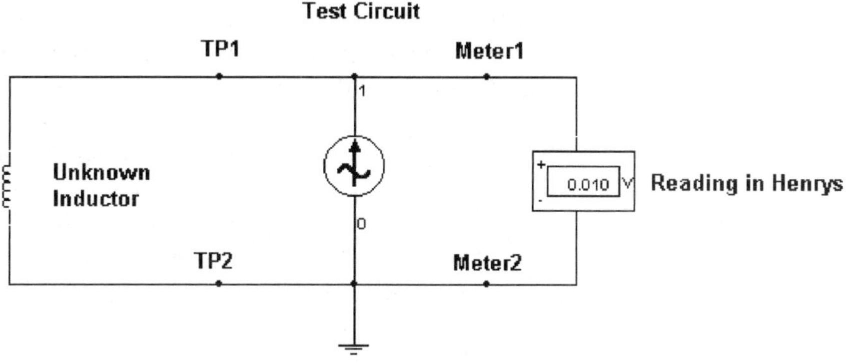


**Figure 12-1** An Inductance Test Circuit using EWB Components

voltmeter to the right is part of the test circuit and will read out the inductance of an unknown coil in units of inductance (**Henrys**). Ignore the "V" on the meter and substitute the symbol for Henrys, **H**, in its place (if the reading on the meter, for example, is 0.180 V, then the inductance is 180 mH). Activate the circuit and determine the inductance of the unknown coil. L = __________ mH.

3. Open circuit file **12-01b**. Determine the inductance of the three coils, $L_1$, $L_2$, and $L_3$. The measured inductance of the three coils is $L_1$ = __________ mH, $L_2$ = __________ H, and $L_3$ = __________ mH.

4. Open circuit file **12-01c**. In this circuit there is a variable inductor. Determine the percentage (%) setting that provides approximately 45 mH. The variable inductor is adjusted to a __________ % setting to achieve a value of 45 mH.

## Activity 12.2: Calculate and Measure Inductance

1. Inductors in series and parallel are similar to resistors in series and parallel. When inductors are in series, the total amount of inductance in the circuit is equal to the sum of the individual inductance values. When inductors are in parallel, they use the same formulas as resistors in parallel to determine total inductance, except that inductance is substituted for resistance.

2. Open circuit file **12-02a**. Calculate the total inductance of the series circuit. The calculated total inductance of the circuit is __________ mH. Activate the circuit and measure the total inductance of the series coils. The measured total inductance of the circuit is __________ mH.

3. Open circuit file **12-02b**. Calculate the total inductance of the parallel circuit. The calculated total inductance of the circuit is __________ mH. Activate the circuit and measure the total inductance of the parallel coils. The measured total inductance of the circuit is __________ mH.

4. Open circuit file **12-02c**. Calculate the total inductance of the series-parallel circuit. The calculated total inductance of the circuit is __________ mH. Activate the circuit and measure the total inductance of the series-parallel coils. The measured total inductance of the circuit is __________ mH.

- ***Troubleshooting Problem:***

5. Open circuit file **12-02d**. This circuit has an unknown coil ($L_1$) and it is necessary to know its value. Use the Inductance Meter and determine the inductance of in-circuit coil $L_1$. It is best to disconnect the coil from the circuit path to avoid false measurements. In this circuit, disconnect one end of the coil and then measure it. If the power supply were grounded, then there would be a conflict between the ground inside the meter circuit and the power supply ground. In that case, take the inductor completely out of circuit to obtain an accurate measurement.

   The inductance of $L_1$ = __________ H.

## Activity 12.3: Calculating and Measuring Inductive Reactance in an AC Circuit

1. As stated previously, inductive reactance is the term used to describe opposition to change in current flow. The symbol for inductive reactance is $\mathbf{X_L}$, and the unit of measurement is the ohm (Ω). Inductive reactance can be calculated, and its effects measured in a circuit, using Ohm's law, Kirchhoff's laws, and the power formulas in a manner similar to resistive circuits. "Real" inductors have resistance (wire resistance) and inductive reactance. Inductive reactance is more commonly taken into consideration in an AC circuit rather than wire resistance. Remember that both parameters are there and sometimes both parameters are important. In Figure 12-2, a circuit is displayed where Ohm's law can be used to calculate inductive reactance.

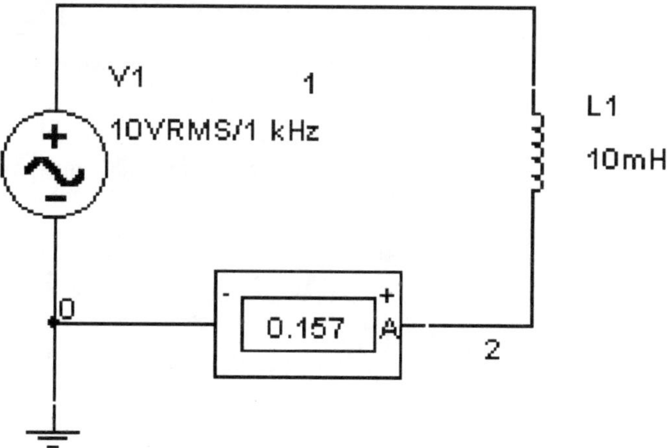


**Figure 12-2** A Circuit where Ohm's Law Can Be Used to Calculate Inductive Reactance

2. Open circuit file **12-03a**. In this circuit, a simple inductive circuit, the applied voltage is given and the ammeter displays the amount of current

flowing in the circuit. Inductive reactance in this case can be solved by using Ohm's law, with one exception: $X_L$ takes the place of R in the formula.

Activate the circuit and determine $X_L$. $X_L$ = __________ kΩ.

3. One big difference between resistive and inductive (reactive) circuits, is the part the frequency of the AC signal plays in the inductive circuit, while having no effect in a simple DC circuit. The formula for $X_L$, (using the inductance factor) is $X_L = 2\pi fL$. This formula has the constant $2\pi$, f standing for the frequency of the AC signal, and L, which represents the amount of inductance of the inductive component. In any AC circuit, inductive reactance has an exact value at a specific frequency of the signal source. If the frequency is changed, then $X_L$ will change as a result.

4. Open circuit file **12-03b**. Using the $X_L$ formula, determine the inductive reactance of the circuit at a frequency of 5 kHz.

   Calculated $X_L$ = __________ kΩ. Activate the circuit and use Ohm's law to verify the calculation. Measured $X_L$ = __________ kΩ. Often there will be slight differences between the measured and the calculated solutions to problems involving inductors and capacitors. The technician has to be able to determine when the discrepancies between measurements and calculations are important and when they are not.

5. Open circuit file **12-03c**. Using the $X_L$ formula, determine the inductive reactance of the circuit at a frequency of 1 kHz.

   Calculated $X_L$ = __________ kΩ. Activate the circuit and use Ohm's law to verify the calculation. Measured $X_L$ = __________ kΩ. How much power does $L_1$ consume? $L_1$ consumes __________ mW of reactive power.

6. Using the knowledge gained from these circuits, it is possible to make a statement concerning $X_L$ and frequency. When frequency goes up, $X_L$ goes __________. When frequency goes down, $X_L$ goes __________. How much power does $L_1$ consume? $L_1$ consumes __________ mW of power.

7. Open circuit file **12-03d**. Calculate the $X_L$ of this circuit and project the value of circuit current. Calculated $X_L$ = __________ Ω.

   Calculated $I_T$ = __________ mA. Activate the circuit and use Ohm's law to verify the calculation. Measured $X_L$ = __________ Ω.

   Measured $I_T$ = __________ mA. How much power does $L_1$ consume?

   $L_1$ consumes __________ mW of power.

- ***Troubleshooting Problem:***

8. Open circuit file **12-03e**. Calculate the $X_L$ of this circuit and project the value of circuit current.

   Calculated $X_L$ = ________ Ω. Calculated $I_T$ = ________ mA. Activate the circuit and verify the calculation concerning $I_T$. Obviously there is a problem; $I_T$ is not correct according to the calculation. What is wrong?

   The problem with the circuit is ________________.

## Activity 12.4: Phase Relationships in Inductive Circuits

1. In a purely inductive circuit, current flow lags voltage by 90°. In other words, the sinusoidal voltage drop across an inductor will lead the current flow through it by 90°. In an electronics circuit with only an inductor in the circuit, it is difficult to display this relationship. It is easy to observe the voltage across the inductor with an oscilloscope (an oscilloscope can only display voltage), but it is not possible to observe the current through the inductor. The way around the problem is to insert a very small resistor in series with the inductor to monitor the circuit current (the current is the same through the resistor as through the inductor in a series circuit). See Figure 12-3 for a typical circuit prepared to measure phase shift. This small resistance will not affect the circuit to any great degree.

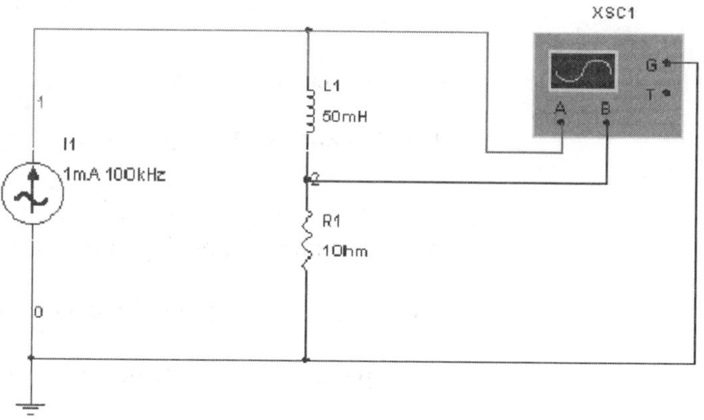


**Figure 12-3** A Test Circuit to Measure Phase Shift through an Inductor

2. Open circuit file **12-04a**. Notice that in this circuit there is a one-ohm resistor in series with the inductor. It is installed to monitor inductor current. The inductor opposes the current change in the sinusoidal signal applied to the circuit and the resistor voltage drop reflects that current and its resultant time lag behind the voltage. Activate the circuit and observe the (almost) 90° phase shift between the two inputs,

applied voltage and inductor current. Channel A is displaying the voltage applied to the inductor, which is at a maximum point to the left of the screen. Channel B reflects the current flowing through the inductor, which is at zero to the left of the screen. This indicates that at the beginning of the sweep (of the scope traces), from left to right, voltage is at a maximum and current is at a minimum; this is a 90° phase difference. Notice that Figure 12-4 displays the 90° phase shift. The top waveform displays the voltage applied to the circuit and the lower waveform displays the current flowing through the inductor. What is the inductive reactance of this circuit?

$X_L$ = __________ Ω. The relationship between resistance and inductive reactance in a circuit influences the circuit in relationship to the ratio of the $X_L$ and R ohmic values. As seen in this circuit, $X_L$ is much larger than R, so R does not have much influence.

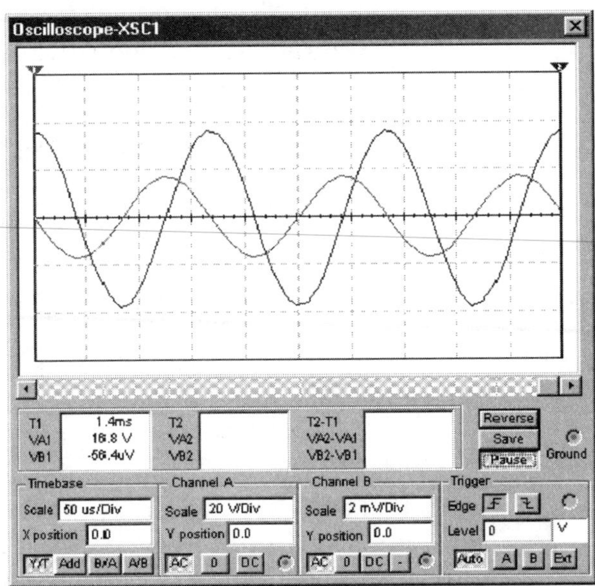


**Figure 12-4** 90° Phase Shift in an Inductive Circuit

3. Open circuit file **12-04b**. Activate the circuit. Is the phase shift 90°?

   Yes ______ or No ______. What is the $X_L$ of the circuit?

   Calculated $X_L$ = __________ Ω. Measured $X_L$ = __________ Ω.

## Activity 12.5: The Q of an Inductor and Power Losses in Coils

1. The symbol for inductor quality is "Q". Real-world inductors not only inject inductance into a circuit, but they also inject resistance, the resistance of the wire that makes up the coil plus some other resistive factors. The term for the resistance plus the other resistive factors is "effective

series resistance" (**ESR**). The formula for Q is $Q = X_L/ESR$. This formula indicates that Q is the ratio between the resistance and the reactance of an inductive circuit. An inductive circuit with both factors shown in the circuit is displayed in Figure 12-5.

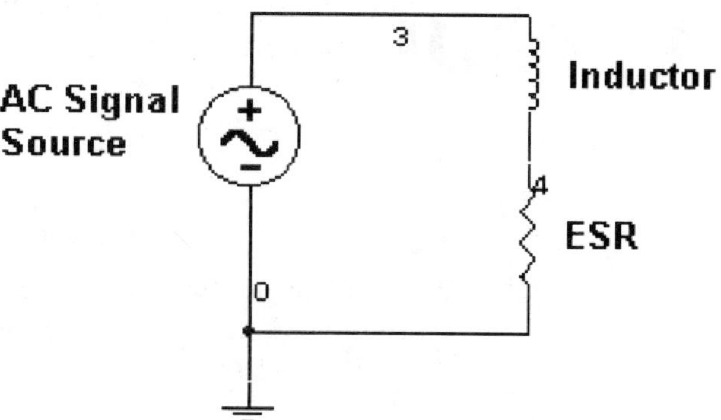


**Figure 12-5** An Inductive Circuit Showing an Inductor and ESR

2. Open circuit file **12-05a**. Calculate the Q of this circuit. The first task in this circuit is to calculate $X_L$. $X_L$ = __________ Ω. Then calculate Q. The Q of the circuit is __________. There is no unit of measurement for Q because it is a ratio of quantities of "like" units.

3. An "ideal" inductor dissipates no power because it does not have resistance. A real-world inductor does dissipate power because there is always some resistance except in the case of superconductors. Power losses in a series inductive circuit are the results of ESR, which is the sum of copper loss (wire resistance), eddy-current loss, hysteresis loss, skin effect loss, and dielectric loss. There are various techniques that are applied to inductors to reduce all of these losses, with varying degrees of effectiveness.

4. What is the amount of power loss in this circuit?

   Power loss = __________ mW. Use one of the power formulas to calculate power loss.

## • *Troubleshooting Problem:*

5. Open circuit file **12-05b**. In this circuit, the current should be about 70.55 mA. Obviously there is something wrong. What is the problem?

   The problem is ______________________________

   ______________________________.

# 13. Resistive-Inductive Circuits

***References***

*Electronics Workbench®, MultiSIM* Version 6

*Electronics Workbench®, MultiSIM* Version 6 Study Guide

**Objectives** After completing this chapter, you should be able to:

- Operate series circuits containing resistors and inductors.
- Operate parallel circuits containing resistors and inductors.
- Operate series-parallel circuits containing resistors and inductors.
- Troubleshoot resistive-inductive circuits.

## Introduction

A **resistive-inductive** (RL) circuit contains resistors and inductors in the form of components that exhibit those parameters. There are also stray resistance and stray inductance throughout a circuit. The electronic activity that takes place in an RL circuit is a combination of the characteristics of a resistive circuit and that of an inductive circuit. Whether the circuit acts more resistive than inductive, or more inductive than resistive, depends on which parameter offers more opposition to current action than the other.

The approach to RL circuits is with some degree of "unknowing." On one hand there is a resistor, which has no voltage/current phase shift, in series with an inductor where the voltage leads the current by 90°. In the last chapter, low value resistors were used to aid in determining phase shift between voltage and current through an inductor; but these resistors did not affect the circuit parameters to any large degree. In the case of large resistances in series with inductances, there will be considerable interaction.

Remember that the current is the same through all the components of a series circuit. The current through the resistor is in-phase with the current through the inductor while the voltage dropped across the inductor is 90° out-of-phase with the current. It is not possible, under these circumstances, to simply sum up all of the individual voltage drops as is done in a simple DC circuit. Instead, a special method of addition called phasor addition will be used along with the Pythagorean theorem.

The opposition to current flow in an RL circuit consists of both resistance and reactance and requires a modified approach to solving circuit parameters. The total opposition to current flow in a circuit containing reactive and resistive components, such as inductors and resistors, is called **impedance (Z)**.

## Activity 13.1: Series Resistive-Inductive (RL) Circuits

1. A series resistive-inductive (RL) circuit contains a resistor and an inductor in series as loads. Figure 13-1 is a series RL circuit as it is presented in EWB. You can look at this circuit as resistive with an inductor added or as inductive with a resistor added.

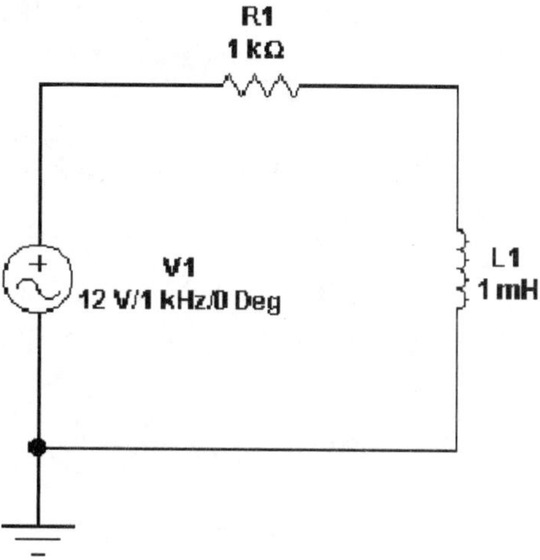


**Figure 13-1** A Series Resistive-Inductive Circuit in EWB

2. Open circuit file **13-01a**. In this series RL circuit, an ammeter is installed to measure series current flowing in the circuit. Activate the circuit and determine current flow from the current meter.

   Measured $I_T$ = __________ mA.

3. Measure the voltage drops across the resistor and the inductor with the DMM. Measured $V_{R1}$ = __________ V and $V_{L1}$ = __________ V. Does the sum of the voltage drops equal the voltage applied to the circuit by the voltage source (12 V)? As you can see, the sum of the voltage drops does not equal the voltage applied, as it should according to Kirchhoff's voltage law.

4. The reason for this problem is that, while the current is the same through both of the series components, the resistor and the inductor, there is a phase difference between the voltage across the resistor and the voltage across the inductor. The amount of phase shift in the circuit depends upon the X-Y vector relationships between the voltage drops. There will be a 90° phase shift between the two voltage drops.

5. The solution to the "voltage drop" situation is found in phasor (addition) mathematics. Using the results of Step 2, draw a proportional phasor on the X-axis representing the voltage drop across the resistor. On the positive Y-axis, draw a proportional phasor representing the voltage drop across the inductor (this represents the 90° phase shift). And, finally, draw a phasor representing the hypotenuse of the two right angles on the plot. This hypotenuse can be measured to get a relative answer for total voltage, or the Pythagorean method can be used to get a more exact answer. This answer should equal the voltage applied to the circuit (12 V). Use the Pythagorean method to obtain an answer.

   Calculated $V_A$ = __________ V.

6. There are also two methods used to obtain a solution for the total opposition (impedance) to current flow problem: the Ohm's law method and the phasor mathematics method. The easiest is the Ohm's law method and it is usually used. In this circuit, measure impedance using Ohm's law.

   Z = __________ Ω.

7. Use the X-Y method and draw a proportional phasor on the positive portion of the X-axis on the plot representing the resistance of the resistor. Calculate the inductive reactance of the inductor and then (on the positive portion of the Y-axis) draw a proportional phasor representing the inductive reactance of the inductor. This represents the 90° phase

   shift. Calculated $X_L$ = __________ Ω. Finally, draw a phasor representing the hypotenuse of the two right angles on the plot. This hypotenuse can be measured to get a relative answer or the Pythagorean method can be used to get a more exact answer. This answer should be close to or equal the opposition to current flow that was previously calculated by Ohm's law. Calculated (the formula for the Pythagorean method is:

   $Z = \sqrt{R^2 + X_L^{\ 2}}$ ). Z = __________ Ω.

8. Open circuit file **13-01b**. Determine circuit impedance (Z) and $V_A$. In this circuit the first step is to solve for inductive reactance ($X_L$) and then

   impedance (Z). Calculated $X_L$ = __________ Ω and Z = __________ Ω. After that, $V_A$ can be determined using Ohm's law. Calculated (Ohm's

   law) $V_A$ = __________ V. Measure the voltage drops across the resistor and the inductor with the DMM and then use the measurements to determine

   $V_A$ using the Pythagorean method. Measured $V_{R1}$ = __________ Ω and

   $V_L$ = __________ Ω.

   Calculated (Pythagorean method) $V_A$ = __________ V.

9. Open circuit file **13-01c**. Determine $X_L$, Z, $V_A$, $V_{R1}$, $V_L$, and $I_T$. In this circuit, again, the first step is to solve for inductive reactance and then impedance. $X_L$ = __________ Ω and Z = __________ Ω. Next, $V_A$ can be determined by measuring the voltage drops across the components and using the Pythagorean method of solution.

   Measured $V_R$ = __________ V and $V_L$ = __________ V. Calculated (Pythagorean method) $V_A$ = __________ V. Finally, $I_T$ can be determined by Ohm's law. Calculated (Ohm's law) $I_T$ = __________ mA.

10. Open circuit file **13-01d**. Determine $X_L$, Z, $I_T$, and $V_A$. $X_L$ = __________ Ω, Z = __________ Ω, $I_T$ = __________ A, and $V_A$ = __________ V. The determination of total power consumed in a series circuit also uses the phasor/Pythagorean method of calculation. The power consumed by the reactive (VARs, volt-amps reactive) component and the resistive (watts) component is placed on each respective axis and total power (VA) is the hypotenuse of the right triangles.

    $P_{R1}$ = __________ W, $P_{L1}$ = __________ VAR, and $P_T$ = __________ VA.

11. Open circuit file **13-01e**. Determine circuit parameters and enter the data in Table 13-1. Verify the $R_1$ calculation with the DMM. The measured value of $R_1$ = __________ Ω.

| | **Current** | **Voltage** | **Power** | **Opposition to Current Flow** |
|---|---|---|---|---|
| $R_1$ | | | W | R = |
| $L_1$ | | | VAR | $X_L$ = |
| Totals | | | VA | Z = |

**Table 13-1** Parameters of a Series RL Circuit

12. Open circuit file **13-01f**. Determine circuit parameters and fill in Table 13-2. Verify the $R_1$ calculation with the DMM.

    The measured value of $R_1$ = __________ Ω.

| | Current | Voltage | Power | Opposition to Current Flow |
|---|---|---|---|---|
| $R_1$ | | | W | R = |
| $L_1$ | | | VAR | $X_L$ = |
| Totals | | | VA | Z = |

**Table 13-2** Parameters of a Series RL Circuit

- ***Troubleshooting Problems:***

13. Open circuit file **13-01g**. There is a problem with this circuit; the coil is smoking and too much current is being drawn from the power source. Determine $X_L$ and Z first and then calculate the proper value of $I_T$. Calculated $X_L$ = __________ Ω, Z = __________ Ω, and $I_T$ = __________ A. Use the DMM, measure the voltage drops, and determine the fault. The problem is ______________________________________________ ______________________________________________.

14. Open circuit file **13-01h**. There is a problem with this circuit; the resistor is overheating and too much current is being drawn from the power source. Determine $X_L$ and Z first and then calculate the proper value of $I_T$. Calculated $X_L$ = __________ Ω, Z = __________ Ω, and $I_T$ = __________ mA. Use the DMM, measure voltage drops and determine the fault. The problem is ______________________________________________.

## Activity 13.2: Parallel Resistive-Inductive (RL) Circuits

1. As we discovered in series DC (resistive) circuits, parallel RL circuits have identical voltage drops across every parallel component in the circuit (see Figure 13-2). The current will divide between the parallel branches according to the amount of opposition (resistance or reactance) that each parallel branch offers to that flow of current. Phase relationships are another matter and differ from series RL circuits. In parallel RL circuits the voltage is the same across all of the parallel components, but there is always a current and voltage phase shift across inductors. The current through inductors is shifted in phase by 90° from the resistors.

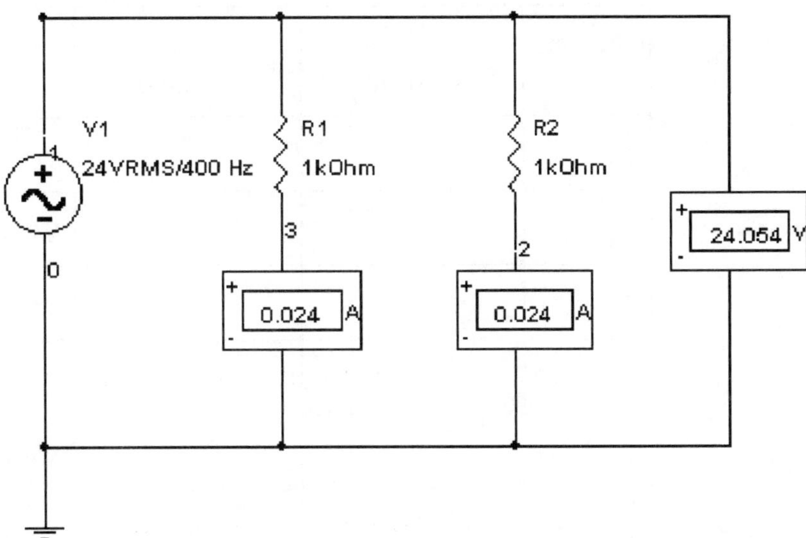


**Figure 13-2** A Parallel RL Circuit in *Electronics Workbench*®

2. Open circuit file **13-02a**. This circuit displays the voltage drop across the parallel components ($M_1$), the branch currents ($M_3$ and $M_4$), and the total circuit current flow ($M_2$). Activate the circuit and total the current flow through the parallel branches. Is the sum of the branch currents equal to $I_T$ as reflected by $M_2$? Obviously not, so we will use phasor addition to calculate $I_T$. Use the phasor method with resistor current reflected on the positive X-axis and inductor current reflected on the negative Y-axis. Then use the Pythagorean method to determine circuit current. Calculated $I_T$ = __________ mA. The calculated current using phasors should equal the current displayed on meter $M_3$. Use Ohm's law and determine the reactance offered by the inductive branch of the circuit ($X_L = V_A/I_L$) and then determine the impedance using Ohm's law ($Z = V_A/I_T$).

   $X_L$ = __________ Ω and Z = __________ Ω.

3. Open circuit file **13-02b**. Determine the circuit parameters and fill in Table 13-3.

| | Current | Voltage | Power | Opposition to Current Flow |
|---|---|---|---|---|
| $R_1$ | | | W | $R_1 =$ |
| $R_2$ | | | W | $R_2 =$ |
| $R_T$ | | | W | $R_T =$ |
| $L_1$ | | | VAR | $X_{L1} =$ |
| $L_2$ | | | VAR | $X_{L2} =$ |
| $L_T$ | | | VAR | $X_{LT} =$ |
| Totals | | | VA | $Z =$ |

**Table 13-3** Parallel RL Circuit Parameters

4. In parallel RL circuits, the Ohm's law method is the easiest method to use to determine impedance when current and voltage parameters are available, but there is a mathematical method using one of the parallel resistance formulas and the Pythagorean theorem shown in Figure 13-3. What is Z using this formula? Z = __________ Ω.

$$Z = \frac{R \times X_L}{\sqrt{R^2 + X_L^{\,2}}}$$

**Figure 13-3** An Alternate Method (Mathemetical) to Determine Impedance in a Parallel RL Circuit

## • *Troubleshooting Problem:*

5. Open circuit file 13-02c. Determine circuit parameters and fill in Table 13-4. The two resistors have the same value of resistance and should have the same amount of current flowing through them. Also, the two inductors, with the same values of inductance and inductive reactance, should have the same amount of current flowing through them. There appear to be two problems in this circuit. Find them. The two problems are

__________________________________________, and

__________________________________________.

| | Current Should Be | Current Is | Voltage Is | Power Should Be | Opposition to Current Flow |
|---|---|---|---|---|---|
| $R_1$ | | | | W | $R_1 =$ |
| $R_2$ | | | | W | $R_2 =$ |
| $R_T$ | | | | W | $R_T =$ |
| $L_1$ | | | | VAR | $X_{L1} =$ |
| $L_2$ | | | | VAR | $X_{L2} =$ |
| $L_T$ | | | | VAR | $X_{LT} =$ |
| Totals | | | | VA | $Z =$ |

**Table 13-4** Parallel RL Circuit Parameters

## Activity 13.3: Series-Parallel Resistive-Inductive (RL) Circuits

1. In a manner similar to series-parallel DC (resistive) circuits, series-parallel RL circuits are a combination of series sections where series-circuit parameters operate and parallel sections where parallel-circuit parameters operate (Figure 13-4). Each parallel RL section has to be simplified until it is minimized, hopefully, into a series RL section. Then further simplification can take place with series components. As with resistive circuits, the process of circuit simplification starts at the farthest point from the power source and proceeds toward the power source.

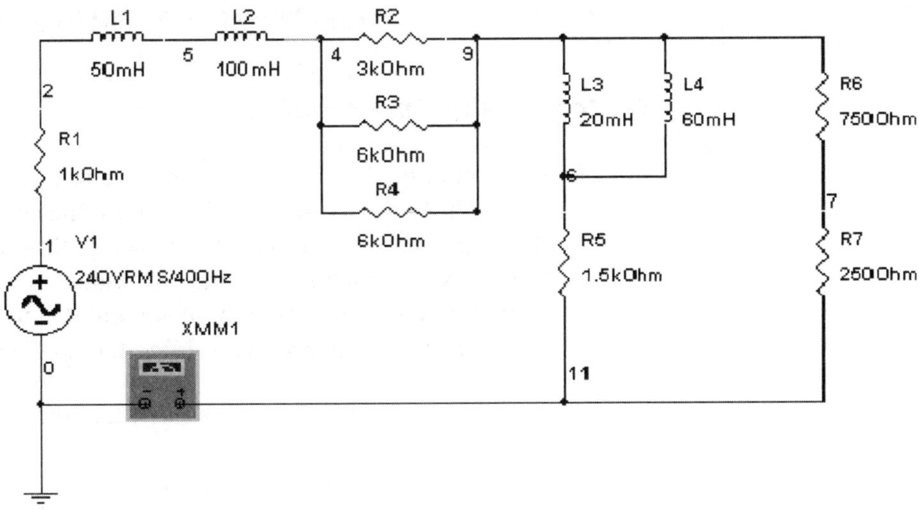


**Figure 13-4** A Series-Parallel RL Circuit in *Electronics Workbench®*

2. Open circuit file **13-03a**. Simplify this circuit. Activate the circuit and record the circuit current. $I_T$ = __________ mA. Combine resistors $R_6$ and $R_7$ into combination resistor $R_{6\text{-}7}$ and resistors $R_1$, $R_2$, $R_3$, and $R_4$ into combination resistor $R_{1\text{-}2\text{-}3\text{-}4}$. Combine inductors $L_1$ and $L_2$ into combination inductor $L_{1\text{-}2}$ and inductors $L_3$ and $L_4$ into combination inductor $L_{3\text{-}4}$.

   $R_{1\text{-}2\text{-}3\text{-}4}$ = __________ Ω, $R_{6\text{-}7}$ = __________ Ω, $L_{1\text{-}2}$ = __________ mH,

   and $L_{3\text{-}4}$ = __________ mH.

3. Open circuit file **13-03b**. Change the value of the resistors and inductors to reflect the combination components of the previous circuit (see Figure 13-4). Activate the circuit and record the circuit current. The current in this circuit should be the same as previously recorded in Step 2.

   $I_T$ = __________ mA.

4. Open circuit file **13-03c**. Simplify this circuit into a single resistor and a single inductor.

   $R_{SIMPLIFIED}$ = __________ kΩ and $L_{SIMPLIFIED}$ = __________ mH.

## ● *Troubleshooting Problems:*

5. Open circuit file **13-03d**. Activate the circuit and record $I_T$.

   $I_T$ = __________ mA. The circuit current is supposed to be 27.0–31.0 mA and it is much higher. In this circuit $L_1$ is defective and it is supposed to be 122 mH. The circuit current is supposed to be 29.24 mA and it is much higher. While looking through the parts bins in the shop, four inductors that appear to be the same type as $L_1$ were found. Measure all four of them and replace $L_1$ with the one that is closest in value to the 120 mH value that $L_1$ is supposed to measure. The coil closest in value is

   L __________. Activate the circuit; what does $I_T$ measure now?

   $I_T$ = __________ mA.

6. The part was replaced and the equipment turned on. Smoke rises and the replacement for $L_1$ is history. Upon continuing the troubleshooting process another bad component is found that is causing $L_1$ to smoke and it is replaced. The equipment has to be operational while a new $L_1$ is on order. Is there any series-parallel combination with the other three coils that could temporarily replace $L_1$ until the correct replacement part arrives and if so, what is the combination? A possible solution would be

   ______________________________. Change the parts and check the circuit.

# Activity 13.4: Pulse Response and L/R Time Constants of RL Circuits

1. So far in our study of RL circuits, circuit operation with sinusoidal signals has been emphasized. Now we are going to look at the action of RL circuits under non-sinusoidal conditions such as with square waves and with the unique situation in DC circuits at turn-on and turn-off time. The operation of an RL circuit under square wave conditions is similar to the turn-on and turn-off situation in a DC circuit where there is a sudden transition from off-to-on or on-to-off. RL components normally react to these sudden changes that take place in a DC circuit, but after that initial reaction, cease to respond to normal DC operating conditions.

2. Open circuit file **13-04a1**. In this circuit, we are going to view the increase in $V_A$ from 0 V to 10 V over several time constants with the scope connected across the resistor. Channel A of the oscilloscope is connected across the resistor ($R_1$).

3. Turn on the circuit and activate the switch to watch the rise in voltage from 0 V to 10 V in 5 TC. Determine the voltage two seconds after the point the rise time starts. The voltage after two seconds = __________ V. You may place the red or blue cursor on the two-second point and read the voltage directly from the related window.

4. Next, open circuit file **13-04a2**. Activate the circuit and view the decrease in $V_L$ from 10 V to 0 V over five time constants. What is the duration of one time constant? 1 TC = __________ seconds. Notice that the scope is set so that one major division is equal to 1.0 second.

5. Once again, the oscilloscope had proven to be a powerful tool to view the inner workings of an electronic circuit. If the $V_R$ and $V_L$ waveforms were displayed simultaneously, the scope would display the two curves crossing each other in a manner similar to the Universal time-constant chart. What is the time duration of five time constants in this circuit?

   5 TC = __________ seconds. What is the voltage after five time constants?

   The voltage after five time constants is __________ V.

6. Open circuit file **13-04b1**. Determine the voltage rise in time constant intervals and enter the data in Table 13-5.

| Time Constant | 1 TC | 2 TC | 3 TC | 4 TC | 5 TC |
|---|---|---|---|---|---|
| Voltage Reading | | | | | |

**Table 13-5** Voltage Rise Time in Time Constant Intervals

7. Open circuit file **13-04b2**. Determine the voltage fall time in time constant intervals and enter the data in Table 13-6.

| **Time Constant** | **1 TC** | **2 TC** | **3 TC** | **4 TC** | **5 TC** |
|---|---|---|---|---|---|
| Voltage Reading | | | | | |

**Table 13-6** Voltage Fall Time in Time Constant Intervals

8. An RL circuit has a predicable rise and fall time when sudden transitions occur in a DC circuit. The response of an RL circuit (or any circuit) to the fast rise and fall time of square waves is called "step response."

9. Open circuit file **13-04c**. Activate the circuit and enlarge the oscilloscope to observe the step response of an RL circuit to a square wave input. Use the pause button on the scope to stop the trace action. The response across the inductor is being compared with the input square wave. The red trace on the oscilloscope is the square wave input and the blue trace is the **differentiated** response of the inductor. This is considered to be a short time constant circuit. When the step occurs in this circuit, the inductor initially drops all of the voltage and then, after a period of five time constants, assumes its normal operating condition in a DC circuit with 0 V voltage drop across the inductor. What would be the duration of five time constants in this circuit? 5 TC = __________ msec.

10. Open circuit file **13-04d**. Activate the circuit and observe the step response in two modes of **integrator** action. The display for the short time constant mode is the type of signal that is usually considered to be an output of an integrator circuit. The display for the long time constant circuit has some integrator action as is observable from the rounding off of the leading and lagging edges of the square wave. The only difference between the two time constant circuits is the size of the resistor. If the resistance value of $R_2$ were made larger, then the integration action would be less obvious. What is the duration of the long and the short time constants in this circuit? The long time constant = __________ μsec and the short time constant = __________ μsec.

## • *Troubleshooting Problem:*

11. Open circuit file **13-04e**. You are developing this circuit for the engineering department at work and the circuit operating specification is for an extremely short time constant resulting in a differentiated pulse that looks a spike as displayed in Figure 13-5 (do not change the function generator or oscilloscope settings). Activate the circuit and observe the

signal on the oscilloscope. There are two choices in this circuit: (1) to increase/decrease the resistance of the resistor, or (2) to increase/decrease the inductance of the inductor. Which would be the least expensive?

Probably to ______________________________________________
would be less expensive.

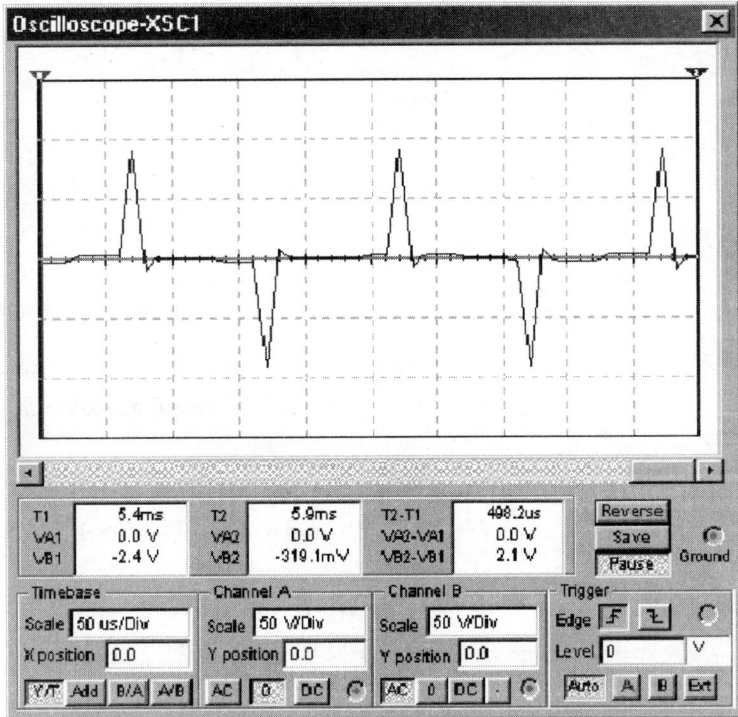


**Figure 13-5** A Differentiated "Spike" Waveform

# 14. Capacitance and Capacitive Reactance

***References***

*Electronics Workbench®, MultiSIM* Version 6

*Electronics Workbench®, MultiSIM* Version 6 Study Guide

**Objectives** After completing this chapter, you should be able to:

- Use an EWB capacitance meter to measure capacitance of a capacitor.
- Calculate and measure capacitance in series, parallel, and series-parallel capacitive circuits.
- Calculate and measure capacitive reactance and impedance in a circuit.
- Determine phase relationships of current and voltage in a capacitive circuit.
- Operate circuits containing capacitance.
- Troubleshoot capacitive circuits.

## Introduction

**Capacitance (C)** is another basic phenomena of electronics and is used in many applications. There are many industrial applications that use capacitance and capacitive characteristics to their advantage. Capacitance is defined as the property, in a circuit, that opposes a change in voltage and has the ability to store a charge in an electrostatic field. A **capacitor** is a component that has capacitance and is also called a condenser. The basis for understanding capacitance is closely related to an understanding of electrostatic charges.

Capacitors have many uses in electronics including being used in the "tuned" circuits found in the fields of radio and television transmission and reception, and as filters in electronic circuits to remove undesired signals and characteristics.

The first definition of capacitance, opposition to a change in voltage, indicates that capacitance is an active device in alternating current circuits where there is a constant change of voltage. In DC applications, capacitors block current flow through them and attempt to keep voltage constant in circuits where there is any change in DC voltage levels.

The second characteristic of a capacitor is its ability to store a charge in an electrostatic field. The charge is stored in the electrostatic field that is built up on the conducting surfaces that constitute the capacitor. In a DC circuit, the electrostatic field charge is built up on the conducting surfaces and remains there until power is removed or there is a change in the DC voltage level. In an AC circuit, the electrostatic field charge is constantly changing since the voltage continuously changes its level as a result of the continually changing current flow. In an AC circuit, capacitors attempt to keep voltage levels constant while simultaneously, the current is varying. As the current tries to change direction and the voltage follows suit, the stored energy in the established electrostatic field attempts to keep the voltage from changing, thus opposing the change in voltage level.

Another concept to understand concerning capacitance and capacitors is "how much opposition they offer to changing voltage levels." The term for the "amount of opposition" to a change in voltage level is "**Capacitive Reactance**." Capacitive reactance (**$X_C$**) is measured in ohms the same as resistance and inductive reactance. For example, $X_C = 10\ \Omega$.

In EWB, the capacitors are found in the **Basic** library on the task bar. There are many types of capacitor choices offered in the basic components in this library and they are all considered to be "**ideal**" components, having the exact value assigned to them. Capacitance meters are commonly available in the electronics lab and are necessary to measure capacitance in and out of circuit during the troubleshooting of electronics equipment. We will discuss one developed as a special EWB circuit.

## Activity 14.1: Testing Capacitors with a Capacitor Test Circuit

1. In Chapter 1, the various types of capacitors were introduced. Now we will study capacitance, capacitors, and capacitive reactance. In its simplest terms, a capacitor is two conducting surfaces separated by an insulator (a dielectric). Next we will look at a simple test circuit.

2. Open circuit file **14-01a**. In this file a simple test circuit checks the charging capabilities of a capacitor. This test circuit is designed with EWB components to test an unknown capacitor and is shown in Figure 14-1. The voltmeter displays the charging voltage, gradually increasing until the capacitor is fully charged. This is only a relative type of test and is similar to the ohmmeter method of testing a capacitor in the field. In the field, using an ohmmeter, the charging capability of a capacitor under test is compared with a known "good" capacitor of equal or similar value. The technician will use this type of test in the field depending on his/her past experience to ascertain whether the capacitor under test is good or bad. Under laboratory conditions a more accurate method of testing, such as with a capacitance tester or a component bridge circuit, is used to determine capacitance. How many seconds does it take to charge the capacitor (to about 8.5 V)? It takes about __________ seconds to charge the

capacitor to 8.5 V. If a capacitor is defective, the technician has to determine the type of defect from experience of typical readings on his/her test equipment. If the meter reads zero volts (in our test circuit), the capacitor is shorted. If the meter reads some voltage and doesn't charge properly, the capacitor is leaky. If the capacitor immediately reads full voltage, the capacitor is open.

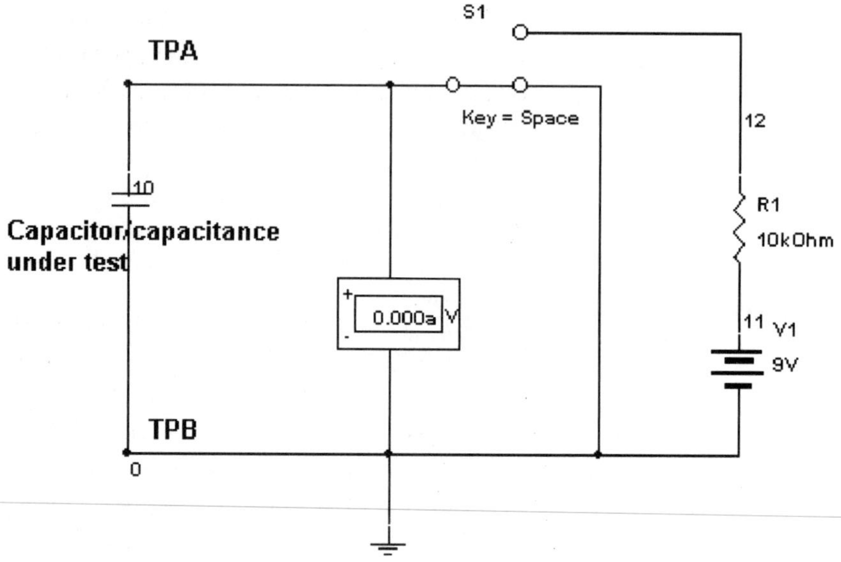


**Figure 14-1** A Capacitance Test Circuit in EWB

3. Open circuit file **14-01b**. Check the four identical capacitors with the test circuit and notice the three capacitors that are defective. $C_1$ is good, $C_2$ is leaky / open / shorted, $C_3$ is leaky / open / shorted, and $C_4$ is leaky / open / shorted (circle the defects for $C_2$, $C_3$, and $C_4$). **Note:** When the screen displays the announcement shown in Figure 14-2, answer "**Yes**" and continue with your test procedure.

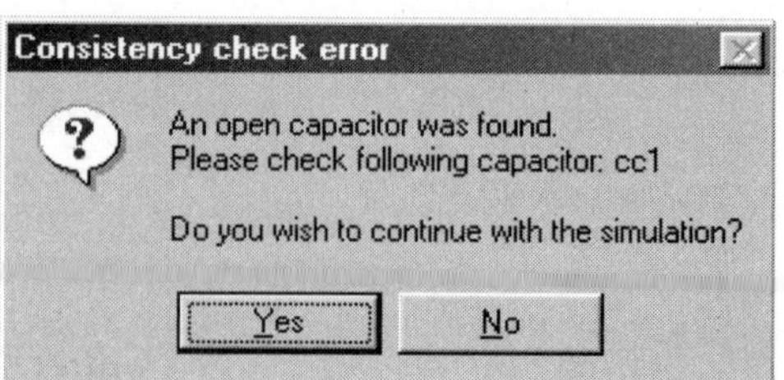


**Figure 14-2** Screen Display of **Consistency Check Error**

## • *Troubleshooting Problems:*

4. One simple test of capacitors is using an ohmmeter to see if the capacitors will charge. This is a pretty rough test, giving little information

except that the capacitor seems to be working properly. It does not give a capacitor value; it simply tells the technician whether the capacitor will charge, if it is shorted, if it is open, or if it is leaky. With experience or with a "known to be good" comparison capacitor, technicians can determine whether a capacitor is charging correctly.

5. Open circuit file **14-01c**. In this circuit there are six capacitors that have to be tested to determine their condition. Three are good (they charge), one is open, one is shorted (0.0000 Ω) and one is leaky (100 Ω). Identify the various conditions and drag the correct title over from the right side of the screen and place it under the correct capacitor.

   $C_1$ is ____________, $C_2$ is ____________, $C_3$ is ____________,

   $C_4$ is ____________, $C_5$ is ____________, and $C_6$ is ____________.

6. How long does it take for a good capacitor such as $C_1$ to charge and reach its fully charged condition? To get a fairly accurate time measurement, place the mouse pointer over the pause button on the upper right of the screen (see Figure 14-2) and when the meter stops increasing in voltage change, left-click the mouse on the pause button. Then read the time interval from the bottom of the screen. It takes about ____________ seconds for $C_1$ to reach a fully charged condition.

## Activity 14.2: Calculating and Measuring Total Capacitance

1. The method used to determine the total capacitance for capacitors in series is calculated in a manner similar to resistors and inductors in parallel, using one of the parallel formulas such as: $C_1 \times C_2/C_1 + C_2$, or one of the other parallel (resistor) formulas. Various circuits are shown in Figure 14-3.

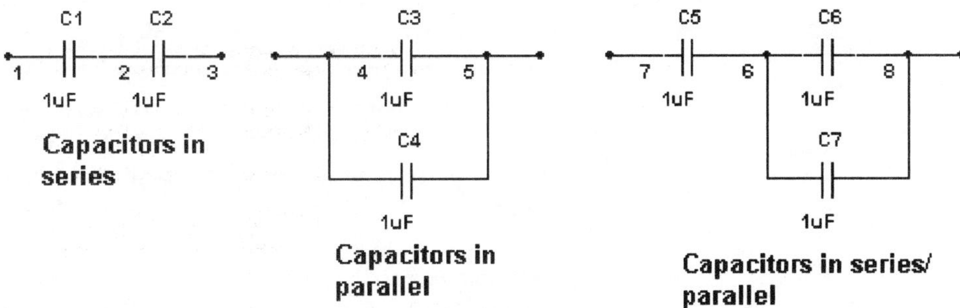


**Figure 14-3** Examples of Capacitive Circuit Configurations

2. The determination of total capacitance for capacitors in parallel is conducted in a manner similar to resistors and inductors in series; the capacitance values add and total capacitance is equal to the sum of all the parallel capacitors.

3. Open circuit file **14-02a**. Calculate the total capacitance of the series circuit.

   The calculated total capacitance of the circuit is __________ µF.

4. Open circuit file **14-02b**. Calculate the total capacitance of the parallel circuit. The calculated total capacitance of the circuit is __________ µF.

5. Open circuit file **14-02c**. Calculate the total capacitance of the series-parallel circuit. The calculated total capacitance of the circuit is

   __________ µF.

- ***Troubleshooting Problem:***

6. Open circuit file **14-02d**. There is a problem in this circuit; the current flow is supposed to be about 4.84 mA and it is 5.5 mA instead. Also, the circuit has a capacitor ($C_1$) whose value is unknown and it is necessary to know its value. Use the ohmmeter function of the DMM to detect the circuit fault. It is best to disconnect the capacitor to measure its characteristics.

   The problem in the circuit is ______________________________.

## Activity 14.3: Calculating and Measuring Capacitive Reactance in an AC Circuit

1. As previously stated, capacitive reactance ($X_L$) is the term used to delineate the amount of opposition to a change in voltage offered by a capacitor with that opposition being contained in the built-up electrostatic field. The unit of measurement for capacitive reactance is the ohm. Capacitive reactance can be calculated; and its effects measured in a circuit using Ohm's law, Kirchhoff's laws, and the power formulas in a manner similar to resistive and inductive circuits. "Real" capacitors not only have capacitive reactance, but they also have some power losses due to the resistance of the conducting surfaces, leakage resistance, and dielectric dissipation. We will ignore these power losses in this study. EWB capacitors are to be considered "ideal" with no power losses. Capacitive reactance is the most important factor regarding capacitors that needs to be considered in a circuit, but remember that the power loss problems are always there too.

2. The easiest way to calculate capacitive reactance in a series circuit is the Ohm's law method; divide the circuit voltage by the circuit current. In Figure 14-4 Ohm's law parameters can be used to calculate capacitive reactance.

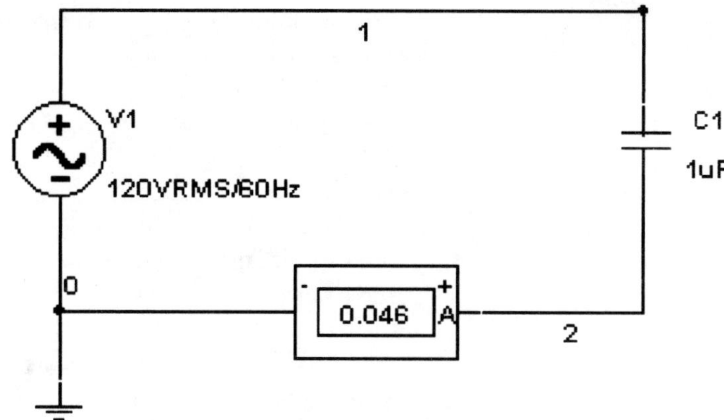


**Figure 14-4** Calculate Capacitive Reactance in a Circuit using Ohm's Law

3. Open circuit file **14-03a**. In this simple capacitive circuit the applied voltage is given and the ammeter displays the amount of current flowing in the circuit. Use the meter readings and Ohm's law to solve for capacitive reactance in this circuit. $X_C$ takes the place of R in the Ohm's law formula and $X_C = V_1/I_T$. Activate the circuit and determine $X_C$.

   Measured $X_C$ = __________ kΩ.

4. One big difference between resistive and capacitive (reactive) circuits is the part that the frequency of the AC signal plays in a reactive circuit while having little or no effect in a simple DC circuit. The formula for $X_C$ (using the capacitance formula), is $X_C = 1/2\pi fC$. In this formula, $2\pi$ is a constant (6.28), f stands for the signal frequency the capacitor is responding too, and C stands for the amount of capacitance of the capacitive component (most commonly in μF). In any AC circuit, capacitive reactance has an exact value at a certain frequency of the signal source and if the frequency is changed, then $X_C$ will change proportionally.

5. Open circuit file **14-03b**. Using the $X_C$ formula, determine the capacitive reactance of the circuit at a frequency of 500 Hz.

   Calculated $X_C$ = __________ kΩ. Activate the circuit and use Ohm's law to verify the calculation. Measured $X_C$ = __________ kΩ.

6. Open circuit file **14-03c**. Using the $X_C$ formula, determine the capacitive reactance of the circuit at a frequency of 1.6 kHz.

   Calculated $X_C$ = __________ kΩ. Activate the circuit and use Ohm's law to verify the calculation. Measured $X_C$ = __________ kΩ. Notice that in this circuit the capacitance is less than 1 μF.

7. Using the knowledge gained from the circuits that you have just worked with, make a statement concerning $X_C$ and frequency.

   When frequency goes up, $X_C$ goes ________. When frequency goes down, $X_C$ goes ________.

8. Open circuit file **14-03d**. Calculate the $X_C$ of this circuit and project the value of circuit current.

   Calculated $X_C$ = ________ Ω. Projected $I_T$ = ________ mA. Activate the circuit and use Ohm's law to verify the calculations.

   Measured $X_C$ = ________ Ω. Measured $I_T$ = ________ mA.

- ***Troubleshooting Problem:***

9. Open circuit file **14-03e**. Calculate the $X_C$ of this circuit and project the value of circuit current flow. Calculated $X_C$ = ________ Ω.

   Calculated $I_T$ = ________ mA. Activate the circuit and verify the calculation concerning $I_T$. Obviously there is a problem; $I_T$ is not correct according to calculations. What is wrong?

   The problem with the circuit is ________________.
   Use the DMM to check the circuit.

## Activity 14.4: Phase Relationships in Capacitive Circuits

1. In a purely capacitive circuit, voltage lags current flow by 90°. In other words, the sinusoidal charging current flow onto the capacitor plates will lead the voltage across the capacitor by 90°. In an electronics circuit with only a capacitor in the circuit, it is difficult to display this relationship. It is easy to observe the voltage across the capacitor with an oscilloscope (an oscilloscope can only display voltage), but it is not possible to observe the current flow to the capacitor. The way to get around the problem is to insert a very small resistor in series with the capacitor to monitor the circuit current (the current is the same through the resistor and the capacitor in a series circuit). Figure 14-5 displays a typical circuit to measure phase shift. This solution is similar to the method used with inductors to determine voltage phase shift.

2. Open circuit file **14-04a**. In this circuit there is a 10-Ω resistor in series with the capacitor that is installed to monitor capacitor current. The capacitor opposes the voltage change in the sinusoidal signal applied to the circuit and the resistor voltage drop reflects the voltage and its resultant lag behind the current.

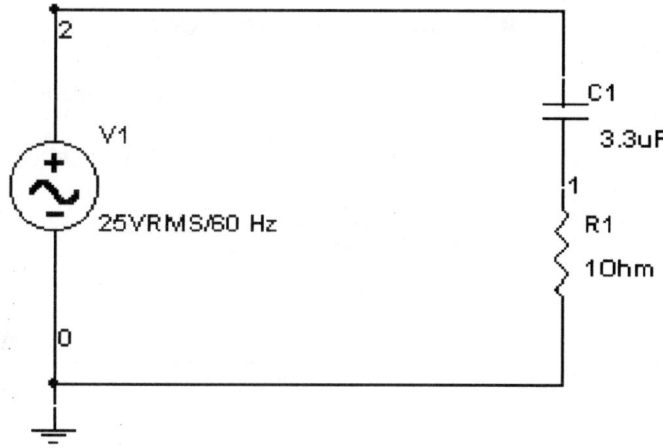


**Figure 14-5** Circuit for Measuring Phase Shift in a Capacitive Circuit

3. Activate the circuit and observe the 90° (almost) phase shift between the two inputs, applied voltage and circuit current. Channel A is displaying the voltage applied to the capacitor, which is at zero at the left of the screen. Channel B reflects the current flowing through the resistor, which is high at the left of the screen when the voltage is at zero. This indicates that at the beginning of the sweep (of the scope traces), from left to right, voltage is at a minimum and current is at a maximum; this is a 90° phase difference. Figure 14-6 shows the phase shift with the top waveform displaying the voltage applied to the circuit and the lower waveform displaying the current flowing through the capacitor.

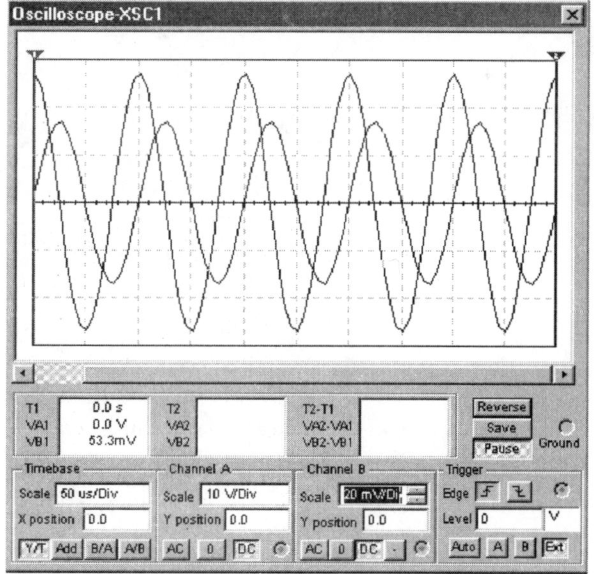


**Figure 14-6** A 90° Phase Shift in a Capacitive Circuit

4. Open circuit file **14-04b**. Activate the circuit. Is the phase shift 90°?

   Yes _____ or No _____. What is the $X_C$ of the circuit? $X_C$ = ___________ Ω.

## • *Troubleshooting Problem:*

5. Open circuit file **14-04c**. Activate the circuit. Is the phase shift 90°?

   Yes ____ or No ____. What is the $X_C$ of the circuit?

   Calculated $X_C$ = __________ Ω.

   What is the problem? The problem is ______________________________

   ______________________________________________________________.

# 15. Resistive-Capacitive Circuits

**References**

*Electronics Workbench®, MultiSIM* Version 6

*Electronics Workbench®, MultiSIM* Version 6 Study Guide

## Objectives

After completing this chapter, you should be able to:

- Operate series circuits containing resistors and capacitors.
- Operate parallel circuits containing resistors and capacitors.
- Operate series-parallel circuits containing resistors and capacitors.
- Determine pulse response and time constants of RC circuits.
- Troubleshoot resistive-capacitive circuits.

## Introduction

A **resistive-capacitive (RC)** circuit contains resistance and capacitance usually in the form of components that exhibit those parameters. The activity that takes place in a circuit of this type is a combination of both characteristics, that of a purely resistive circuit and that of a purely capacitive circuit. Whether the circuit acts more resistive than capacitive, or more capacitive than resistive, depends on which parameter offers more opposition to current flow than the other.

When looking at RC circuits, observe that this type of circuit is a combination of resistors and capacitors where there is no voltage and current phase shift concerning the resistor, yet there is a 90° voltage versus current phase shift concerning the capacitor. It is not possible, under these circumstances, to sum up the individual voltage drops as it is done in a simple DC circuit. Instead, the same mathematical tools of phasor addition along with the Pythagorean theorem that were developed in the chapters on inductance will be used.

## Activity 15.1: Series Resistive-Capacitive (RC) Circuits

1. The series resistive-capacitive (RC) circuit contains a resistor and a capacitor in series as loads. In Figure 15-1 a series RC circuit as it is presented in EWB *MultiSIM®* is displayed. This series circuit contains an AC signal source, a capacitor, and a resistor.

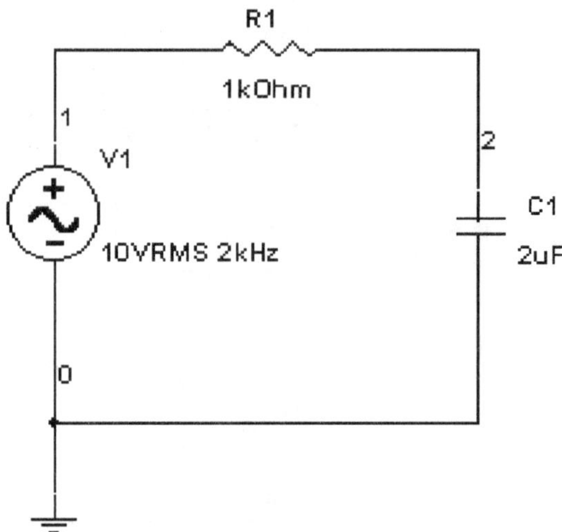


**Figure 15-1** A Series RC Capacitive Circuit in EWB

2. Open circuit file **15-01a**. In this series RC circuit, an ammeter is installed to measure series current flowing in the circuit. Remember, according to Ohm's law, current is the same everywhere throughout a series circuit. Also, resistive and reactive components drop voltage in proportion to the amount of their resistance or reactance.

3. Activate the circuit and determine current flow.

   Measured $I_T$ = __________ mA. Measure the voltage drop across both the resistor and the capacitor. $V_{R1}$ = __________ V and $V_{C1}$ = __________ V. Does the sum of the voltage drops equal the voltage applied to the circuit by the voltage source? The apparent sum of the voltage drops (does / does not) equal the voltage applied which seems to be contrary to the precepts of Kirchhoff's voltage law.

4. The primary reason for this problem lies in the fact that, while the current is the same through both of the series components, the resistor and the capacitor, there is a phase difference between the voltage drop across the resistor and the voltage drop across the capacitor. The amount of this phase shift depends on the vector relationship between the resistance and the reactance, which both impede current flow. If the circuit were purely resistive, there would be no phase difference and if the circuit were purely reactive there would be a 90° phase shift between voltage and current. The amount of phase shift is relative to the proportions of resistance and capacitive reactance present in the circuit.

5. The solution is found in phasor (addition) mathematics. For a solution to the voltage problem, draw a proportional phasor on the positive X-axis of a plot that represents the voltage drop across the resistor and on the negative Y-axis proportional phasor representing the voltage drop across the capacitor (this represents the 90° phase shift). Then draw a phasor

representing the hypotenuse of the two right angles on the plot. This hypotenuse can be measured to get a relative answer or the Pythagorean method can be used to get a more exact answer. The calculated answer should be close to the reading on the current meter.

The calculated voltage $V_A$ = __________ V.

6. To determine opposition to current flow in this circuit, draw a proportional phasor on the X-axis of a plot representing the resistance of the resistor. Then, calculate $X_C$ and on the negative portion of the Y-axis, draw a proportional phasor representing the calculated capacitive reactance of the capacitor (this represents the 90° phase shift).

   Calculated $X_C$ = __________ Ω. Finally, draw a phasor representing the hypotenuse of the two right angles on the X-Y plot. This hypotenuse can be measured to get a relative answer or the Pythagorean method can be used to get a more exact answer. This answer should be close to the total opposition to current flow that was previously calculated using Ohm's law.

   Calculated Z = __________ Ω. Remember that impedance (Z) is comprised of all reactance and resistance in a circuit, the total opposition offered to voltage/current change in the circuit. Verify your calculation using Ohm's law to determine impedance.

   Measured Z = __________ Ω.

7. Open circuit file **15-01b**. Activate the circuit and determine circuit impedance (Z), $X_C$, and $V_A$. In this circuit, you need to calculate capacitive reactance ($X_C$) first and then impedance (Z).

   Calculated $X_C$ = __________ Ω and Z = __________ Ω. After that, $V_A$ can be determined using Ohm's law.

   Measured (Ohm's law) $V_A$ = __________ V. Measure the voltage drops across the resistor and the capacitor with the DMM and then use the measurements to determine $V_A$ using the Pythagorean method.

   Calculated (Pythagorean method) $V_A$ = __________ V.

8. Open circuit file **15-01c**. Activate the circuit and determine $I_T$, Z, and $V_A$. In this circuit, calculate capacitive reactance first and then impedance.

   $X_C$ = __________ Ω and Z = __________ Ω. Next, $V_A$ can be determined by measuring the voltage drops across the components and then using the Pythagorean method of solution.

   Measured $V_R$ = __________ V and $V_C$ = __________ V.

   Calculated (Pythagorean method) $V_A$ = __________ V. Finally, $I_T$ can be determined by Ohm's law using the calculated circuit voltage ($V_A$) and

   impedance ($I_T = V_A/Z$). Calculated $I_T$ = __________ mA.

9. Open circuit file **15-01d**. Activate the circuit and determine $X_C$, Z, $I_T$, and $V_A$.

   $X_C$ = ________ Ω, Z = ________ Ω, $I_T$ = ________ A, and

   $V_A$ = ________ V.

10. Open circuit file **15-01e**. Activate the circuit and determine circuit parameters to fill in Table 15-1. Verify the $R_1$ calculation with the DMM.

    The measured value of $R_1$ = ________ Ω.

| | Current | Voltage | Opposition to Current Flow |
|---|---|---|---|
| $R_1$ | | | R = |
| $C_1$ | | | $X_C$ = |
| Totals | | | Z = |

**Table 15-1** Parameters of a Series RC Circuit

11. Open circuit file **15-01f**. Activate the circuit and determine circuit parameters to fill in Table 15-2. Verify the $R_1$ calculation with the DMM.

    The measured value of $R_1$ = ________ Ω.

| | Current | Voltage | Opposition to Current Flow |
|---|---|---|---|
| $R_1$ | | | R = |
| $C_1$ | | | $X_C$ = |
| Totals | | | Z = |

**Table 15-2** Parameters of a Series RC Circuit

## • *Troubleshooting Problems:*

12. Open circuit file **15-01g**. Calculate the values of $X_C$ and Z first and then $I_T$.

    Calculated $X_C$ = ________ Ω, Z = ________ Ω, and $I_T$ = ________ mA. Activate the circuit and note that the current, which should be about 20.86 mA, is too low. Use the DMM to determine the circuit fault. The

    problem is ________________________________.

13. Open circuit file **15-01h**. There is a problem with this circuit; the resistor is smoking, and too much current is being drawn from the power source. Determine the values of $X_C$ and Z and then calculate the "should be" value of $I_T$.

    Calculated $X_C$ = ________ Ω, Z = ________ Ω, and $I_T$ = ________ A. Activate the circuit, use the DMM to measure voltage drops, and determine the fault.

    The problem is ________________________________________.

## Activity 15.2: Parallel Resistive-Capacitive (RC) Circuits

1. As in parallel DC circuits, parallel RC circuits have identical voltage drops across every parallel component in the circuit (see Figure 15-2). The current will divide between the parallel branches according to the amount of opposition (resistance or capacitive reactance) that each parallel branch offers to the flow of current. Phase relationships are another matter. Differing from series RC circuits, parallel RC circuits have the same voltage across all of the parallel components. There is always a current and voltage phase shift across capacitors, so the current through capacitors must be shifted in phase relationships from the voltage by 90°.

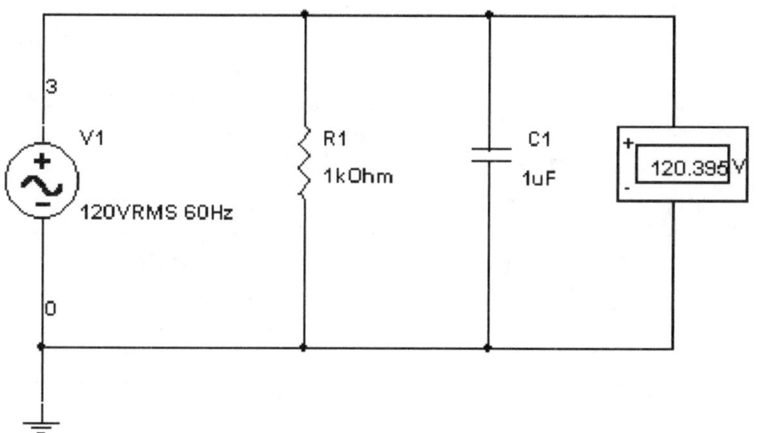


**Figure 15-2** A Parallel RC Circuit in *Electronics Workbench*®

2. Open circuit file **15-02a**. This circuit displays the voltage drop across the parallel components and the current flow through the parallel branches. Total the current flow through the parallel branches. Activate the circuit and determine the arithmetic sum of the currents of the parallel branches.

   The sum of the current in the parallel branches = ________ mA. Is the total of the branch currents equal to $I_T$ as reflected by $M_3$?

   Measured $I_T$ = ________ mA.

3. Use phasor addition to calculate $I_T$, with resistor current reflected on the positive portion of the X-axis and capacitor current reflected on the negative portion of the Y-axis. Then use the Pythagorean method to determine circuit current. The calculated current should equal the current displayed on meter $M_3$.

   Calculated $I_T$ = __________ mA. Now determine the reactance offered by the capacitive branch of the circuit. $X_C$ = __________ Ω. Determine the impedance of the circuit using Ohm's law method ($Z = V_A/I_T$).

   Z = __________ Ω.

4. In a parallel RC circuit, the Ohm's law method is the easiest method to determine impedance when current and voltage parameters are available, but there is a mathematical method using the parallel resistance formulas and the Pythagorean theorem (see Figure 15-3).

   What is the value of Z using this formula? Z = __________ Ω.

$$Z = \frac{R \times X_C}{\sqrt{R^2 + X_C{}^2}}$$

**Figure 15-3** An Alternate Method (Mathematical) to Determine Impedance in a Parallel RC Circuit

5. Open circuit file **15-02b**. Activate the circuit to determine circuit parameters and fill in Table 15-3.

| | Current | Voltage | Opposition to Current Flow |
|---|---|---|---|
| $R_1$ | | | $R_1$ = |
| $R_2$ | | | $R_2$ = |
| $R_T$ | | | $R_T$ = |
| $C_1$ | | | $X_{C1}$ = |
| $C_2$ | | | $X_{C2}$ = |
| $C_T$ | | | $X_{CT}$ = |
| Totals | | | Z = |

**Table 15-3** Parallel RC Circuit Parameters

### • *Troubleshooting Problem:*

6. Open circuit file **15-02c**. Activate the circuit to determine circuit parameters and fill in Table 15-4. The resistors have the same value of resistance and should have the same amount of current flowing through them. Also, the two capacitors have the same value of capacitance and thus, the same value of capacitive reactance and should also have the same amount of current flowing through them. There appear to be two defects in this circuit.

   The problems are ____________________________________.

| | Current Should Be | Current Is | Voltage Is | Opposition to Current Flow |
|---|---|---|---|---|
| $R_1$ | | | | $R_1 =$ |
| $R_2$ | | | | $R_2 =$ |
| $R_T$ | | | | $R_T =$ |
| $C_1$ | | | | $X_{C1} =$ |
| $C_2$ | | | | $X_{C2} =$ |
| $C_T$ | | | | $X_{CT} =$ |
| Totals | | | | $Z =$ |

**Table 15-4** Parallel RC Circuit Parameters

## Activity 15.3: Series-Parallel Resistive-Capacitive (RC) Circuits

1. A series-parallel RC circuit is a combination of series RC sections and parallel RC sections where series and parallel circuit parameters operate as seen in Figure 15-4. Each individual parallel RC section has to be simplified until it is minimized, hopefully, into simple series R and C sections. Then, further simplification can take place with all of the series components.

2. Open circuit file **15-03a**. Activate the circuit and record the circuit current.

   $I_T =$ __________ mA.

3. Calculate the necessary combinations to simplify the circuit. Combine resistors $R_6$ and $R_7$ into combination resistor $R_{6\text{-}7}$ and resistors $R_1$, $R_2$, $R_3$, and $R_4$ into combination resistor $R_{1\text{-}2\text{-}3\text{-}4}$. Combine capacitors $C_1$ and $C_2$ into combination capacitor $C_{1\text{-}2}$ and capacitors $C_3$ and $C_4$ into combination capacitor $C_{3\text{-}4}$.

$R_{1\text{-}2\text{-}3\text{-}4}$ = __________ Ω, $R_{6\text{-}7}$ = __________ Ω, $C_{1\text{-}2}$ = __________ μF, and

$C_{3\text{-}4}$ = __________ μF.

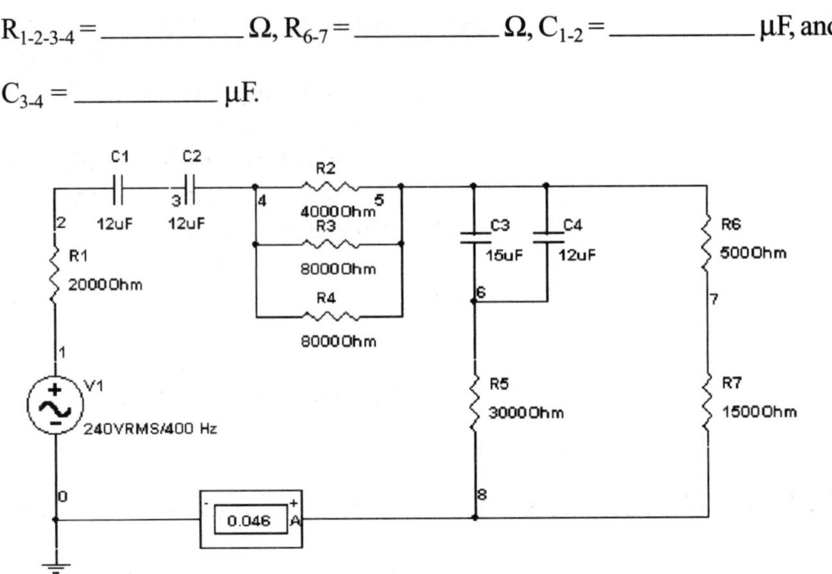


**Figure 15-4** A Series-Parallel RC Circuit in *Electronics Workbench®*

4. Open circuit file **15-03b**. Change the value of the resistors and capacitors to reflect the various combinations of components in the previous circuit (15-03a). Activate the new circuit and record the circuit current.

   $I_T$ = __________ mA. The current should be the same as in the previous circuit if your calculations were correct and correctly placed in the circuit.

5. Open circuit file **15-03c**. Simplify this circuit into a single resistor and a single capacitor.

   $R_{SIMPLIFIED}$ = __________ Ω and $C_{SIMPLIFIED}$ = __________ μF.

## • *Troubleshooting Problems:*

6. Open circuit file **15-03d**. Activate the circuit and record $I_T$.

   $I_T$ = __________ mA. The circuit current is supposed to be 360–370 mA

   and it is much higher. Determine which capacitor is defective. C _____ is defective. In looking through the parts bins in the shop, four capacitors were located that are of the same type. Replace $C_1$ with the one that is closest in value.

   The closest in value is C _____. Activate the circuit.

   What does $I_T$ measure now? $I_T$ = __________ mA.

7. After replacing the part and turning on the equipment, a "poof" occurs, the end comes out of the capacitor, tinfoil shoots across the room, and the replacement capacitor is history. In continuing the troubleshooting

process another bad component in the equipment is found that caused the replacement capacitor to go bad and it is replaced. Now it is necessary to install a temporary part to get the equipment temporarily operational while parts are on order. Is there any series-parallel combination with the other capacitors making it possible to temporarily replace $C_1$ (plus or minus 25% tolerance) until the correct replacement part arrives and if so, what is the combination?

8. After checking out the available parts situation, the best solution is to ______________________________. Change the parts and check the circuit.

## Activity 15.4: Pulse Response and Time Constants of RC Circuits

1. Up to this point our study of RC circuits has been under sinusoidal signal conditions. Next, we will study the operation of RC circuits under non-sinusoidal conditions, primarily square waves, but also the unique situation in DC circuits at turn-on and turn-off time. The operation of an RC circuit under square wave conditions is similar to the turn-on and turn-off situation in a DC circuit where there is a sudden transition from off-to-on or on-to-off. RC circuits do react to these sudden changes that take place in a DC circuit, but after the initial reaction, the capacitor performs the function of blocking DC current flow only.

2. Open circuit file **15-04a**. Activate the circuit and view the increase in $V_A$ from 0 V to 10 V over five time constants. What is the duration of one TC?

   One TC = __________ msec. After viewing rise time, toggle the circuit switch down (space bar) to observe fall time (you may have to toggle the on/off switch). Fall time is a reverse image of rise time.

3. Determine the voltage at the end of forty milliseconds (four major divisions).

   The voltage after 40 msec = __________ V. How many time constants is 30 msec? 30 msec = __________ TC. Place the red or blue cursor on the 30 msec point and read the voltage directly from the related window. The voltage as indicated by the "red cursor" window at the 30 msec point is

   __________ V.

4. Once again, the oscilloscope has proven to be a powerful tool allowing us to view the inner workings of an electronic circuit. When displaying the $V_R$ and $V_C$ waveforms simultaneously, the two curves would cross each other in a manner similar to the Universal time-constant chart. What is the time duration of five time constants in this circuit?

5 TC = __________ msecs. What is the voltage after five time constants?

The voltage after five time constants is __________ V.

5. Open circuit file **15-04b**. Determine the voltage rise in time constant intervals and enter the data in Table 15-5. You will need to use the pause button on the oscilloscope panel to stop trace action and to be able to view rise time.

| **Time Constant** | **1 TC** | **2 TC** | **3 TC** | **4 TC** | **5 TC** |
|---|---|---|---|---|---|
| Voltage Reading | | | | | |

**Table 15-5** Voltage Rise Time in Time Constant Intervals

6. RC circuits have a predicable rise and fall time when sudden transitions occur in a DC circuit. The next subject under study is the response of an RC circuit to the fast rise and fall time of square waves (this is called "step response"). The two types of capacitor action, dependent upon capacitor placement in the circuit, are called integration and differentiation. Look at Figure 15-5 to observe the two types of circuits.

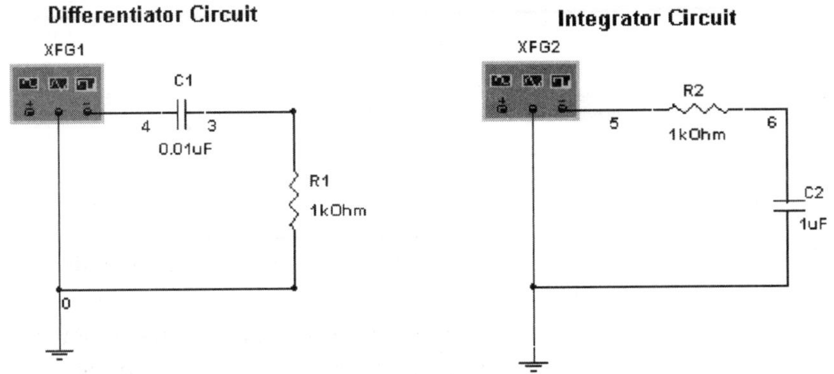


**Figure 15-5** Circuits Showing where Integration and Differentiation Take Place

7. Open circuit file **15-04c**. Activate the circuit and observe the step response of an RC circuit to a square wave. The response across the capacitor is being compared with the input square wave. The upper red trace is the square wave input and the lower blue trace is the "integrated" response of the capacitor to the signal. Calculate the duration of five time constants in this circuit?

5 TC = __________ msec. Does the Channel B signal ever reach the five time constant point in the display? Yes _____ No _____.

8. Open circuit file **15-04d**. Activate this integrator circuit and observe the step response in two modes of integrator action. The display for the long time constant mode is the type of signal that is usually considered to be an output of an integrator circuit. The display for the short time constant circuit has some integrator action, as can be seen from the rounding off of the leading and lagging edges of the square wave. The only difference between the two time constant circuits is in the size of the resistor. If the resistance value of $R_2$ were made larger, then the integration action would be less obvious. Calculate the duration of one time constant (1 TC) for each of the two time constant options in this circuit?

   The short time constant option is __________ msec and the long time constant option is __________ msec.

9. Open circuit file **15-04e**. Activate this differentiator circuit and observe the step response of a capacitor in series with the square wave input to the circuit. As with the integrator, the response (sharpness) of the differentiated square wave is dependent on the values of the series capacitor and the load resistor. If the resistance value of $R_1$ were made smaller, then the differentiation action would be sharper (the differentiated signal would be of shorter duration). If the capacitance value of the capacitor were made smaller, the differentiated signal duration would be less. What is the duration of one time constant in this RC circuit?

   1 TC = __________ μsec.

## • *Troubleshooting Problem:*

10. Open circuit file **15-04f**. Activate the circuit. In assembling this circuit for the engineering department at work, the requirement for an extremely short time constant resulting in a differentiated pulse that looks like (the way it is set up on the oscilloscope) a spike became a necessity. It is now necessary to choose between two options: (1) to increase/decrease the resistance of the resistor, or (2) to increase/decrease the capacitance of the capacitor. Which would be the least expensive?

    Probably to ______________________________
    would be less expensive.

# 16. Resistive-Inductive-Capacitive Circuits

***References***

*Electronics Workbench®, MultiSIM* Version 6

*Electronics Workbench®, MultiSIM* Version 6 Study Guide

**Objectives** After completing this chapter, you should be able to:

- Operate series circuits containing resistors, inductors, and capacitors.
- Operate parallel circuits containing resistors, inductors, and capacitors.
- Operate series-parallel circuits containing resistors, inductors, and capacitors.
- Determine resonant frequencies of RLC circuits.
- Troubleshoot resistive-inductive-capacitive circuits.

## Introduction

**Resistive-inductive-capacitive (RLC)** circuits combine resistors, inductors, and capacitors in various circuit configurations. The activities that take place in a circuit of this type are a combination of all three characteristics: inductance, capacitance, and resistance. The determination of whether the circuit is primarily resistive, inductive, or capacitive depends on the relationships between the three parameters. If one of the three characteristics offers more or less opposition to current action than the other two, the characteristics of the circuit are accordingly modified and the circuit tends to be resistive, inductive, or capacitive depending on which of the three is dominant.

When analyzing RLC circuits, we discover that these circuits are various combinations of resistors, inductors, and capacitors having aiding and opposing interactions within the circuit. There is no voltage and current phase shift with the resistive elements, yet there are 90° voltage and current phase shifts with the capacitors and inductors. And, those phase shifts are in opposition to one another. It is important to remember that the current flow is the same everywhere in a series circuit: through the resistor, the inductor, and the capacitor. While the voltage and current through the resistor is in phase; the current lags the voltage by 90° through the inductor, and the current leads the voltage by 90° through the capacitor.

The parallel and series-parallel circuits offer even more complexity. To solve for circuit parameters with these more complex circuits, use the same methods presented in earlier chapters, the methods of phasor addition along with the Pythagorean theorem, Ohm's law, and Kirchhoff's voltage and current laws.

## Activity 16.1: Series Resistive-Inductive-Capacitive (RLC) Circuits

1. The series **resistive-inductive-capacitive** (RLC) circuit contains resistors, inductors, and capacitors connected in series as loads. In Figure 16-1 a series RLC circuit as it is presented in EWB *MultiSIM*® is displayed. This series circuit contains an AC signal source, a resistor, an inductor, and a capacitor.

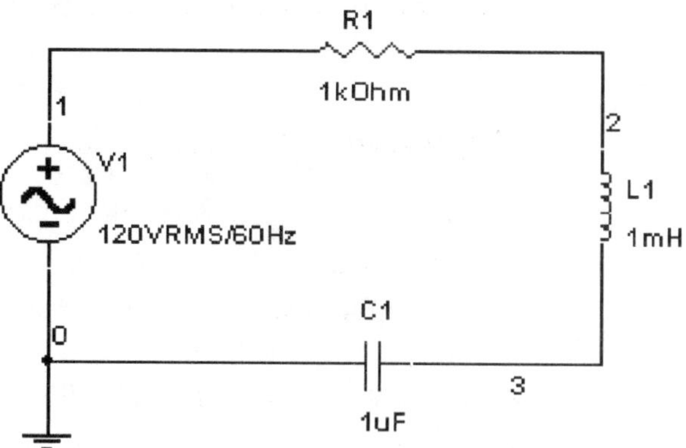


**Figure 16-1** A Series RLC Circuit in EWB

2. Open circuit file **16-01a**. In this series RLC circuit an ammeter is installed to measure series current flowing in the circuit. Remember, according to Ohm's law, current flow is the same everywhere in a series circuit. Also, the resistive and reactive components drop voltages in proportion to their resistance or reactance. Activate the circuit and determine current flow.

   $I_T$ = __________ mA. Measure the voltage drops across the resistor, the inductor, and the capacitor. $V_{R1}$ = __________ V, $V_{L1}$ = __________ V, and $V_{C1}$ = __________ V. Does the sum of the voltage drops equal the voltage applied to the circuit by the voltage source? The sum of the voltage drops (does / does not) equal the voltage applied to the circuit. As has been determined in the study of reactive circuits, there are definite interactions and differences between resistive and reactive circuits.

3. In this type of circuit the phase shifts between the voltage and current of the reactive components are to be expected. The first factor to consider at this point is the differences in the phase shift interactions between the inductors and the capacitors. On one hand, the phase shift has current

leading voltage (capacitive) and on the other hand, current lags the voltage (inductive). Ultimately, the amount of phase shift in either direction will depend upon the degree of opposition offered to the circuit by the inductive reactance and the capacitive reactance. The reactance that is dominant, inductive or capacitive, will establish the characteristic flavor of the circuit. Then in conjunction with resistance as to whether the circuit will act as a RL or a RC circuit. See Figure 16-2 for the formula to solve this type of complex problem without using phasor mathematics. What is the impedance of this circuit using this method? First, calculate $X_L$ and $X_C$.

Calculated $X_L$ = __________ Ω and $X_C$ = __________ Ω. Calculated (impedance formula method) Z = __________ Ω.

$$Z = \sqrt{R^2 + (X_L - X_C)^2}$$

**Figure 16-2** The Formula Used to Solve for Impedance in Series RLC Circuits

4. The Ohm's law formulas are still the easiest to use to determine circuit factors such as impedance, current, or voltage. Determine the amount of opposition (impedance) to current flow in this circuit according to Ohm's law ($Z = V_A/I_T$).

   Calculated Z = __________ Ω.

5. The solution set to the reactance and impedance in this circuit is found in phasor mathematics and the Pythagorean theorem. To solve for circuit impedance, start by drawing a proportional phasor on the X-axis of a plot, representing the resistance of the resistor. On the negative portion of the Y-axis draw a proportional phasor representing the capacitive reactance of the capacitor (this represents one of the 90° phase shifts). On the positive portion of the Y-axis draw a proportional phasor representing the inductive reactance of the inductor (this represents the other 90° phase shift). Next, algebraically sum up the phasors on the Y-axis, subtracting $X_C$ from $X_L$,with dominance being established by the higher value (absolute value) of reactance. Whichever absolute value is larger indicates whether the circuit is ultimately RC or RL. Finally draw a phasor representing the hypotenuse of the two right angles on the plot. This hypotenuse can be measured to get a relative answer or the Pythagorean method can be used to get a more exact answer. What is the impedance of the circuit according to this method?

   Calculated (phasor and Pythagorean method) Z = __________ Ω.

6. Open circuit file **16-01b**. Determine circuit impedance (Z), $X_C$, $X_L$, and $V_A$. In this circuit the first step is to measure the voltage drops.

$V_R$ = ____________V, $V_L$ = ____________V, and $V_C$ = ____________V. Use these voltage drops and the circuit current measurement to calculate inductive reactance ($X_L$), capacitive reactance ($X_C$), and then impedance (Z).

Calculated $X_L$ = __________ Ω, $X_C$ = __________ Ω and

Z = __________ Ω. After that, $V_A$ can be determined using Ohm's law.

Calculated (Ohm's law) $V_A$ = __________ V. Measure the voltage drops across the resistor and the capacitor with the DMM and then use the measurements to determine $V_A$ using the Pythagorean method.

Calculated (Pythagorean method) $V_A$ = __________ V.

7. Open circuit file **16-01c**. Determine $I_T$, Z, and $V_A$. In this circuit, there are several directions that can be taken. Your choice might be to solve for inductive and capacitive reactance and impedance and then measure the voltage drops to determine applied voltage with the Pythagorean method. Or, secondly to calculate circuit current by first measuring the voltage drop across the resistor and then, calculating $I_T$, followed by voltage drop measurements across the reactive components and finally, by calculating reactance using the known current flow. Choose the best method.

   Calculated $X_L$ = __________ Ω, $X_C$ = __________ Ω, and Z = __________ Ω. Next, $V_A$ can be determined by measuring the voltage drops across the components and then using the Pythagorean method of solution.

   Measured $V_R$ = __________ V, $V_L$ = __________ V, and $V_C$ = __________ V.

   Calculated $V_A$ = __________ V. $I_T$ can be determined by Ohm's law.

   Calculated $I_T$ = __________ mA.

8. Open circuit file **16-01d**. Activate the circuit and enter parameters in Table 16-1. Measure $V_R$, $V_L$, and $V_C$.

   Measured $V_R$ = __________ V, $V_L$ = __________ V and $V_C$ = __________ V.

| | **Current** | **Voltage** | **Opposition to Current Flow** |
|---|---|---|---|
| $R_1$ | | | R = |
| $L_1$ | | | $X_L$ = |
| $C_1$ | | | $X_C$ = |
| Totals | $I_T$ = | $V_A$ = | Z = |

**Table 16-1** RLC Series Circuit Data

9. Open circuit file **16-01e**. Determine circuit parameters and enter your data into Table 16-2. Verify the $R_1$ calculation with the DMM.

The measured value of $R_1$ = __________ Ω.

| | Current | Voltage | Opposition to Current Flow |
|---|---|---|---|
| $R_1$ | | | R = |
| $L_1$ | | | $X_L$ = |
| $C_1$ | | | $X_C$ = |
| Totals | $I_T$ = | $V_A$ = | Z = |

**Table 16-2** Parameters of a Series RLC Circuit

10. Open circuit file **16-01f**. Determine circuit parameters and fill in Table 16-3. Verify the $R_1$ calculation with the DMM.

The measured value of $R_1$ = __________ Ω. Remember that the meter connected across $R_1$ will slightly affect the overall measurement. To get a more accurate measurement of the resistor, temporarily disconnect the meter.

| | Current | Voltage | Opposition to Current Flow |
|---|---|---|---|
| $R_1$ | | | R = |
| $L_1$ | | | $X_L$ = |
| $C_1$ | | | $X_C$ = |
| Totals | $I_T$ = | $V_A$ = | Z = |

**Table 16-3** Parameters of a Series RLC Circuit

## ***Troubleshooting Problems:***

11. Open circuit file **16-01g**. There is a problem with this circuit, the current reading is indecisive. Calculate the values of $X_L$, $X_C$, and Z and then the "should be" value of $I_T$.

Calculated $X_L$ = __________ Ω, $X_C$ = __________ Ω, Z = __________ Ω,

and $I_T$ = __________ A. Determine the circuit fault. The problem is

______________________________________________.

12. Open circuit file **16-01h**. There is a problem with this circuit. The current is less than half of its proper value. Determine $X_L$, $X_C$, and Z and then calculate the proper value of $I_T$.

    Calculated $X_L$ = __________ Ω, $X_C$ = __________ , Z = __________ Ω,

    and $I_T$ = __________ mA. Use the test equipment to determine the circuit

    fault. The problem is ______________________________.

## Activity 16.2: Parallel Resistive-Inductive-Capacitive (RLC) Circuits

1. Parallel RLC circuits have identical voltage drops across every parallel component in the circuit (see Figure 16-3) just like the other parallel circuits we have studied. The current will divide between the parallel branches according to the amount of opposition (resistance or reactance) that each parallel branch offers to the flow of current. Phase relationships are another matter. Reactance in a circuit always causes phase shift of current and voltage. As a result, the current through the reactances is phase shifted from the voltage by 90°. This is similar to the RLC series circuits; the reactance that dominates in conjunction with the resistance sets the characteristics of the circuit.

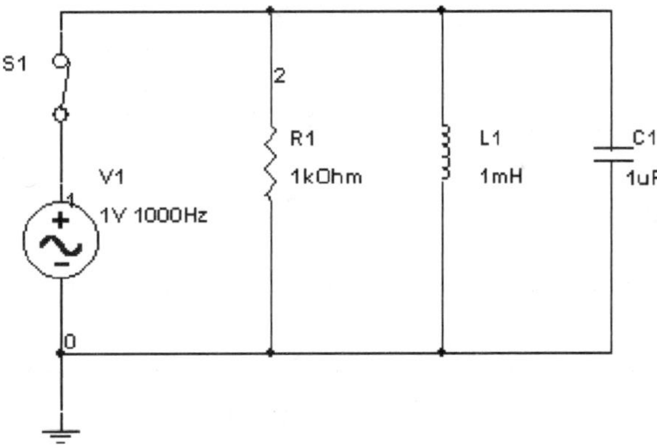


**Figure 16-3** A Parallel RLC Circuit in *Electronics Workbench®*

2. Open circuit file **16-02a**. This circuit demonstrates the voltage drops across the parallel components and the current flow through each component. What is the sum of current flow through the parallel branches?

   The sum of the branch currents = __________ mA. Is the total of the branch currents equal to $I_T$ as reflected by $M_3$? Obviously not.

3. Now use phasor addition to calculate $I_T$, with resistor current reflected on the positive portion of the X-axis, inductor current on the positive portion of the Y-axis, and capacitor current reflected on the negative portion of

the Y-axis. Then use the phasor and Pythagorean method to determine circuit current. The calculated current should equal the current displayed on meter $M_3$.

Calculated $I_T$ = __________ mA. Determine the inductive and capacitive reactance offered by their respective branches of the circuit.

$X_L$ = __________ Ω and $X_C$ = __________ Ω. Determine the impedance of the circuit by the Ohm's law method. Measured Z = __________ Ω.

4. Open circuit file **16-02b**. Determine the circuit parameters and fill in Table 16-4.

| | **Current** | **Voltage** | **Opposition to Current Flow** |
|---|---|---|---|
| $R_1$ | | | $R_1$ = |
| $R_2$ | | | $R_2$ = |
| $R_T$ | | | $R_T$ = |
| $L_1$ | | | $X_{L1}$ = |
| $L_2$ | | | $X_{L2}$ = |
| $L_T$ | | | $X_{LT}$ = |
| $C_1$ | | | $X_{C1}$ = |
| $C_2$ | | | $X_{C2}$ = |
| $C_T$ | | | $X_{CT}$ = |
| Totals | | | Z = |

**Table 16-4** Parallel RLC Circuit Parameters

## ● *Troubleshooting Problem:*

5. Open circuit file **16-02c**. Determine the circuit parameters and then fill in Table 16-5. The resistors have the same value of resistance and should have the same amount of current flowing through them. The two capacitors have the same value of capacitance and should have the same amount of current flowing through them. And, finally, the two inductors also have the same value of inductance and should have the same current flowing

through them. There appear to be two problems in this circuit. Find them.

The problems are ____________________________________.

| | Current Should Be | Current Is | Voltage Is | Opposition to Current Flow |
|---|---|---|---|---|
| $R_1$ | | | | $R_1 =$ |
| $R_2$ | | | | $R_2 =$ |
| $R_T$ | | | | $R_T =$ |
| $L_1$ | | | | $X_{L1} =$ |
| $L_2$ | | | | $X_{L2} =$ |
| $L_T$ | | | | $X_{LT} =$ |
| $C_1$ | | | | $X_{C1} =$ |
| $C_2$ | | | | $X_{C2} =$ |
| $C_T$ | | | | $X_{CT} =$ |
| Totals | | | | $Z =$ |

**Table 16-5** Parallel RLC Circuit Parameters

## Activity 16.3: Series-Parallel Resistive-Inductive-Capacitive (RLC) Circuits

1. As in any other type of series-parallel circuit, series-parallel RLC circuits are a combination of series sections where series-circuit parameters operate and parallel sections where parallel-circuit parameters operate (see Figure 16-4). The first objective in analyzing these types of circuits is to simplify the circuit as much as possible by combining individual components into single entities, joining series components, calculating parallel combinations, and finally reaching a point where no further simplification is possible.

2. Open circuit file **16-03a**. Simplify this circuit. Calculate the necessary combinations to simplify the circuit. Combine resistors $R_4$ and $R_5$ into combination resistor $R_{4\text{-}5}$ and resistors $R_1$, $R_2$, and $R_3$ into combination resistor $R_{1\text{-}2\text{-}3}$. Combine capacitors $C_1$ and $C_2$ into combination capacitor

$C_{1-2}$ and capacitors $C_3$ and $C_4$ into combination capacitor $C_{3-4}$. Finally, combine inductors $L_1$ and $L_2$ into combination inductor $L_{1-2}$ and inductors $L_3$ and $L_4$ into combination inductor $L_{3-4}$.

The values of $R_{1-2-3}$ = ________ Ω, $R_{4-5}$ = ________ Ω,

$L_{1-2}$ = ________ mH, $L_{3-4}$ = ________ mH, $C_{1-2}$ = ________ μF,

and $C_{3-4}$ = ________ μF.

Activate the circuit and record the circuit current. $I_T$ = ________ mA.

3. Open circuit file **16-03b**. Change the value of the resistors and capacitors to reflect the simplification of the series and parallel sections of the previous circuit. Activate the circuit and record the circuit current that should be the same as the previous circuit if the combinations are correctly calculated.

   $I_T$ = ________ mA.

4. Open circuit file **16-03c**. Simplify this circuit into a single resistor, a single inductor, and a single capacitor.

   $R_{SIMPLIFIED}$ = ________ Ω, $L_{SIMPLIFIED}$ = ________ Ω, and

   $C_{SIMPLIFIED}$ = ________ Ω. Calculate $X_L$, $X_C$, Z, and $I_T$.

   Calculated $X_L$ = ________ Ω, $X_C$ = ________ Ω, Z = ________ Ω,

   and $I_T$ = ________ mA.

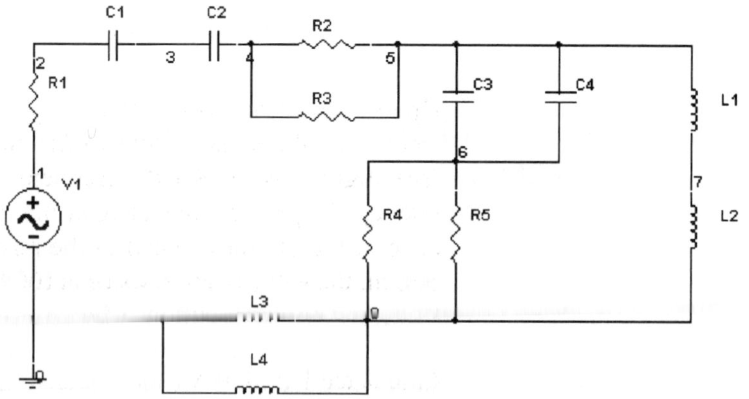

**Figure 16-4** A Series-Parallel RLC Circuit in *Electronics Workbench®*

## • *Troubleshooting Problem:*

5. Open circuit file **16-03d**. This circuit has three problems, one shorted resistor, one open inductor, and one leaky capacitor. Find the defective parts and replace them with the spare parts above the circuit. The

defective components are R ____________, L ____________, and C ____________. Use the replacement components to the right of the circuit and change the values of the replacement components to the rated value of each defective component that is being replaced. Activate the circuit with the new components installed and record $I_T$.

$I_T$ = ____________ mA. If you have replaced the defective components and changed their values to the correct value, circuit current should be 1.547 mA.

## Activity 16.4: Series and Parallel Resonant Circuits

1. In a series RLC circuit at **resonance**, the impedance of the circuit is equal to the resistance of the circuit only. The circuit impedance is at its minimum point (R only) with inductive reactance and capacitive reactance completely canceling each other out as can be shown on an X-Y plot. In other words, at the point of resonance and at a specific frequency, the inductive reactance is equal to the capacitive reactance and the only opposition to current in the circuit is resistance. The voltage drops across the inductor and the capacitor are equal and the current is at a maximum point. The frequency where this condition occurs is called the **resonant frequency** ($f_r$). In every series AC circuit containing inductors and capacitors, there is a resonant frequency. The important point to consider is that at resonance, the voltage drop across the inductor is equal to the voltage drop across the capacitor. The formula for resonant frequency is:

$$f_r = \frac{0.159}{\sqrt{LC}}$$

2. Open circuit file **16-04a**. In this series RLC circuit, the formula for the frequency of resonance tells us that the resonant frequency is 5035.5 Hz (see Figure 16-5). Set the frequency of the signal source to that frequency. As previously stated, at the frequency of resonance the impedance of the circuit is equal to the resistance of the circuit alone. In this circuit, the value of the resistor is 1000 Ω. Calculate circuit current at resonance.

   Calculated $I_T$ at $f_r = V_A/R_1$ = ____________ mA. Activate the circuit and verify your current calculation. Measured $I_T$ at $f_r$ = ____________ mA.

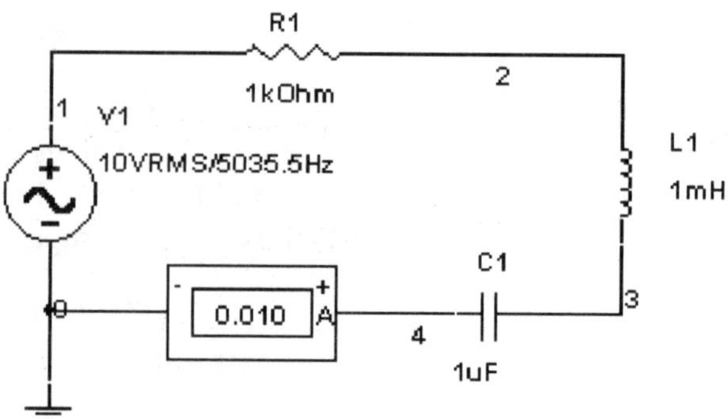


**Figure 16-5** A Series RLC Circuit at the Resonant Frequency

3. Open circuit file **16-04b**. What is the resonant frequency of this circuit and what is the circuit current at that point?

   Calculated $f_r$ = __________ kHz and $I_T$ = __________ mA. Adjust the frequency of the signal source to $f_r$ and activate the circuit. The actual circuit current flow should be close to the calculated value.

   Measured $f_r$ = __________ kHz and $I_T$ = __________ mA.

4. In many aspects, and somewhat to be expected, the parameters for a parallel RLC circuit operate in a manner opposite to a series RLC circuit. Circuit impedance is at its maximum value rather than its minimum. At the resonant frequency the circuit is resistive because inductive reactance and capacitive reactance have cancelled each other out. This causes circuit current to be at a minimum value. In a parallel RLC circuit at its resonant point (see Figure 16-6) the currents in the reactive branches are equal in value (and 180° out of phase with each other). Notice that circuit current is equal to the resistive branch current.

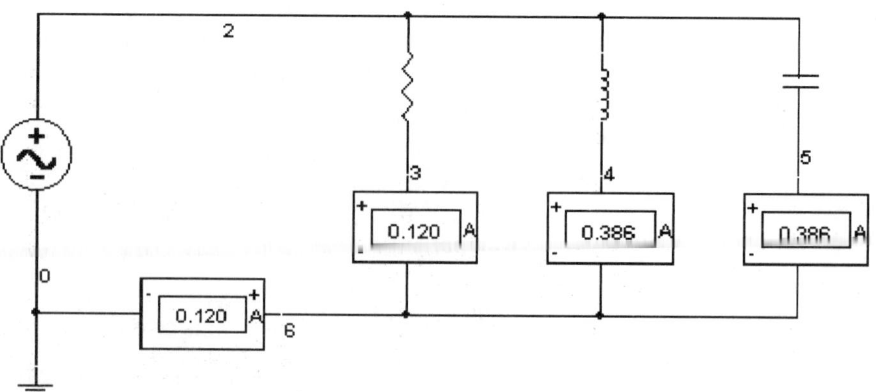


**Figure 16-6** A Parallel RLC Circuit at the Resonant Frequency

5. Open circuit file **16-04c**. In this parallel RLC circuit, the formula for the frequency of resonance (usually the same formula that is used with series

circuits) tells us that the resonant frequency is 1124.3 Hz. Set the frequency of the signal source to that frequency. As previously stated, at the frequency of resonance the impedance of this parallel circuit is equal to the resistance of the circuit alone and is at its maximum point. In this circuit, the value of the resistor is 3000 Ω. Calculate circuit current ($I_T = V_A/R_1$) at resonance.

Calculated $I_T$ at $f_r = V_A/R_1 =$ ___________ mA. Activate the circuit and

verify the current calculation. Measured $I_T$ at $f_r =$ ___________ mA. Notice that the currents are not quite equal in the reactive branches. To get closer to total cancellation of the reactive branches of the circuit, the frequency of the signal source needs to be adjusted closer to a cancellation point. Upon "tweaking" the frequency setting of the signal generator, a generator frequency of 1110 Hz is found to be better. When the circuit is first activated, it is necessary to allow the circuit to stabilize and to reach the point of minimum circuit current. This period of stabilization could take several minutes.

6. Open circuit file **16-04d**. What is the resonant frequency of this circuit and what is the approximate circuit current at that point?

   Calculated $f_r =$ ___________ kHz and $I_T =$ ___________ mA. Adjust the frequency of the signal source to $f_r$ and activate the circuit. The actual circuit current flow should be close to the calculated value.

   Measured $f_r =$ ___________ kHz and $I_T =$ ___________ mA.

## • *Troubleshooting Problems:*

7. Open circuit file **16-04e**. This circuit is not operating properly. The circuit current should be 200 mA and it isn't. What is the problem with the circuit?

   The problem is ___________________________________________.

   What is the calculated value of $f_r$? $f_r =$ ___________ Hz. Set the signal source to the calculated frequency. Activate the circuit.

   What is $I_T$? $I_T =$ ___________ mA.

8. Open circuit file **16-04f**. This circuit does not appear to be operating at its resonant frequency. Calculate the resonant frequency of this circuit and the circuit current at that point?

   Calculated $f_r =$ ___________ Hz and $I_T =$ ___________ mA. Adjust the frequency of the signal source to a value giving the lowest circuit current.

   The adjusted frequency is ___________ Hz. Adjusted $I_T =$ ___________ mA.

# 17. Transformer Circuits

**References**

*Electronics Workbench®, MultiSIM* Version 6

*Electronics Workbench®, MultiSIM* Version 6 Study Guide

**Objectives** After completing this chapter, you should be able to:

- Operate and analyze circuits containing transformers.
- Use step-up and step-down transformers.
- Troubleshoot transformer circuits.

## Introduction

Transformers are standard components in electronics and are usually very small for printed circuit board mounting and similar electronics applications. At the same time, transformers used in industry can weigh tons and be so large cranes and heavy-duty construction equipment are required to install them. In the first type of application the transformer would handle low voltages and currents (or at least low currents). In the larger application, the transformer may be used for voltage distribution, controlling voltages in the hundreds of kilovolts and currents in the thousands of amperes. In almost any location where there is electrical power and electronic circuits, there are transformers. They are one of the most versatile components available for changing AC voltages and currents. Technicians and maintenance personnel must understand transformers and transformer circuits of all types. They have to be able to troubleshoot and analyze circuits and equipment containing transformers.

## Activity 17.1: Transformer Step-Up and Step-Down Circuits

1. **Transformers** operate on the principle of mutual inductance, a magnetic linkage that uses electromagnetic principles to transfer energy from one coil circuit to an adjacent coil circuit. In EWB the transformers are very basic, performing necessary functions to provide transformer action in circuits, but not much beyond that. Figure 17-1 displays types of transformer devices used in EWB *MultiSIM®* Version 6.

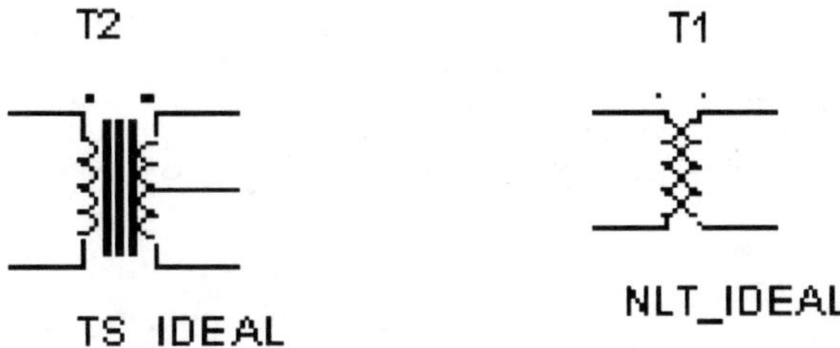


**Figure 17-1** Transformers used in *Electronics Workbench*®

2. The **basic transformer** is just that—basic. It provides all of the necessary functions of a transformer, voltage step up, voltage step down, and voltage isolation. When discussing the step-up and step-down properties of transformers we specifically refer to the step-up or step-down of voltages and not current. In a transformer circuit, current changes in a proportional ratio that is the opposite of the voltage ratio. For example, across the transformer circuit, from primary to secondary, if voltage steps up, then current steps down and if voltage steps down, current steps up.

3. Open circuit file **17-01a**. The default for the **TS_IDEAL** (basic) transformer in EWB is a voltage step down at a 2:1 ratio. The voltage of the primary is twice that of the secondary. Activate the circuit and observe the step-down of the primary voltage from 120 V to a secondary voltage of 60 V. Also observe that on the secondary side of this transformer there is a center tap. This center tap is a point of connection where the secondary voltage can be halved. This is a very necessary feature of transformers that are used in power supply circuits in electronics equipment. Verify the primary and secondary voltages with the DMM.

   Measured primary voltage = ________ V and secondary voltage ________ V. What is the voltage from the ground to the secondary center tap?

   $V_{TAP}$ = __________ V.

4. Open circuit file **17-01b**. Measure the primary and secondary voltages with the DMM.

   Measured $V_{PRIM}$ = __________ V and $V_{SEC}$ = __________ V. What is the primary to secondary turns ratio?

   The turns ratio is __________ to __________.

5. Open circuit file **17-01c**. In this circuit the secondary center tap is going to be checked out. The transformer is set for a 2:1 step-down. The secondary voltage from top to bottom should be one half of the primary voltage. The voltage from the center tap to either end of the secondary winding should be one half of the secondary voltage. The secondary

voltage is ________ V, the center tap to the top connection is ________ V, and the center tap to the bottom connection is ________ V.

6. As mentioned previously, whatever the voltage does in the secondary is reflected in an opposite manner by the current. If the voltage of the secondary steps up by a factor of two, then the current capabilities decrease and are cut in half in this case. If the secondary voltage steps down to one half of the primary voltage, then the secondary current drain capabilities steps up by two. This is modified by a number of factors:
   a. there is always trickle current in the primary that is always present for low levels of current usage by a load,
   b. there is always a power loss (primarily core losses) from primary to secondary (usually less than 5%),
   c. the current drain of the secondary is always reflected back to the primary current drain and the step-up, step-down ratio, and
   d. the power of the primary is equal to the power of the primary excluding core losses.

- ***Troubleshooting Problem:***

7. Open circuit file **17-01d**. In this circuit, the 25:1 step-down transformer has been replaced with a replacement part from the manufacturer. Something is wrong, the load resistor is smoking. The problem is ________________________________.

## Activity 17.2: Transformer Buck/Boost Circuits

1. If a transformer has more than one secondary, the secondary windings can be connected together, resulting in an algebraic summation process. The secondary voltages can add or subtract their voltages from each other depending on the polarity of the series connection. In EWB, the top terminals of the primary and secondary windings have the same polarity as indicated by the polarity dots at the tops of the windings. If the bottom terminal of one secondary is connected to the top terminal of another secondary, the resulting voltage will be positive. If the bottom terminal of one secondary is connected to the bottom terminal of another secondary, then the voltages subtract. If the top terminals are connected together, the voltages also subtract. This method of connecting secondary windings is called a **buck/boost** connection. For example, if one secondary has a voltage of 10 V and another secondary has a voltage of 3 V, then in the boost connection the output voltage would be 13 V and in the buck connection the output voltage would be 7 V.

2. In EWB, the transformers have a single secondary winding. To simulate multiple secondary windings, the primaries of multiple transformers have to be connected to the same voltage with the same polarity of connection (see Figure 17-2). Open circuit file **17-02a**. In this circuit, using the TS_IDEAL default transformers with a 2 to 1 step-down ratio, the transformers are connected in a boost format. Measure the voltage of the series secondary connections.

The voltage $V_{SEC}$ = __________ V.

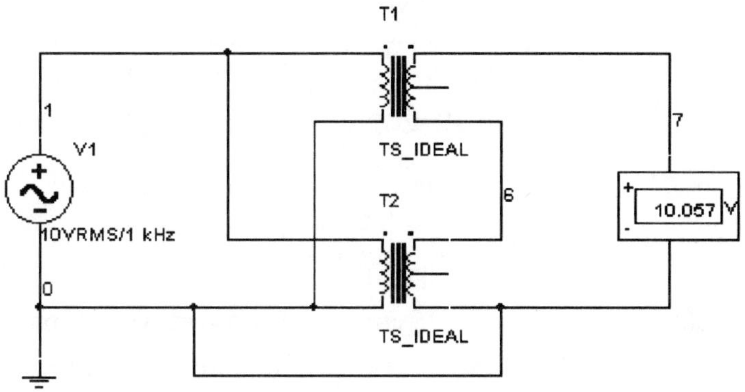


**Figure 17-2** Transformer Boost Connection

3. Open circuit file **17-02b**. In this circuit, the transformers are connected in a buck format (see Figure 17-3). Measure the voltage of the series secondary connections.

The voltage $V_{SEC}$ = __________ V.

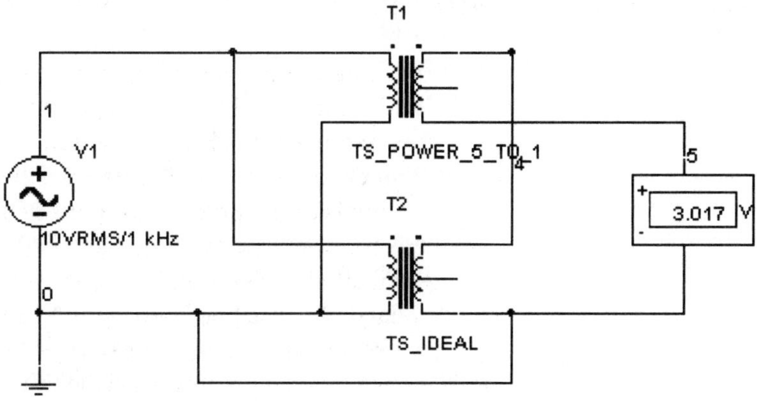


**Figure 17-3** Transformer Buck Connection

# 18. Passive Filter Circuits

***References***

*Electronics Workbench®, MultiSIM* Version 6

*Electronics Workbench®, MultiSIM* Version 6 Study Guide

**Objectives** After completing this chapter, you should be able to:

- Operate and analyze passive filter circuits.
- Use the EWB Bode plotter to analyze circuits.
- Classify passive filters.
- Determine response curves for passive filters.
- Determine cutoff frequencies for filter circuits.
- Analyze coupling and decoupling circuits.
- Troubleshoot transformer circuits.

## Introduction

All reactive components are sensitive to the frequencies of changing signals from signal and power sources as well as AC signals developed in electronic circuits. Filters are circuits that use the "frequency sensitive" characteristics of reactive components. Some filters will pass certain frequencies, others will block them, some will only pass a band of frequencies, and others will block that band. There are many uses for filters in electronics circuits with lack of imagination being the only limitation concerning the application of these circuits.

Filter circuits are categorized in a general manner by their resonance and non-resonance capabilities. Another way that filters are categorized is by the frequency range of their application. Names of filter circuits often reflect their application such as low-pass, high-pass, band-pass, bandstop (notch), and so on.

A special unit of virtual test instrument found only in EWB *MultiSIM®* known as the **Bode plotter** (pronounced bow-dee) can be used to test virtual filter circuits. This virtual (theoretical) instrument is only found on the computer; there will not be one in the electronics lab. The **spectrum analyzer**, in conjunction with an oscilloscope, comes closest to doing the same job as a Bode plotter among "real" test instruments. Historically, there have also been some specialized test instruments, with built-in oscilloscopes, that were used to do similar tasks as the spectrum analyzer, for example, in television bandpass alignment and in specialized military applications.

## Activity 18.1: The EWB Bode Plotter

1. The EWB Bode plotter is used to plot or graph the frequency response of a circuit (see Figure 18-1). This X-Y display represents the voltage level (gain) of the signal being measured on the Y-axis. The display on the Y-axis is in reference to the signal frequency which is plotted on the X-axis. This virtual instrument is typically used to determine frequency response and phase shift in AC circuits.

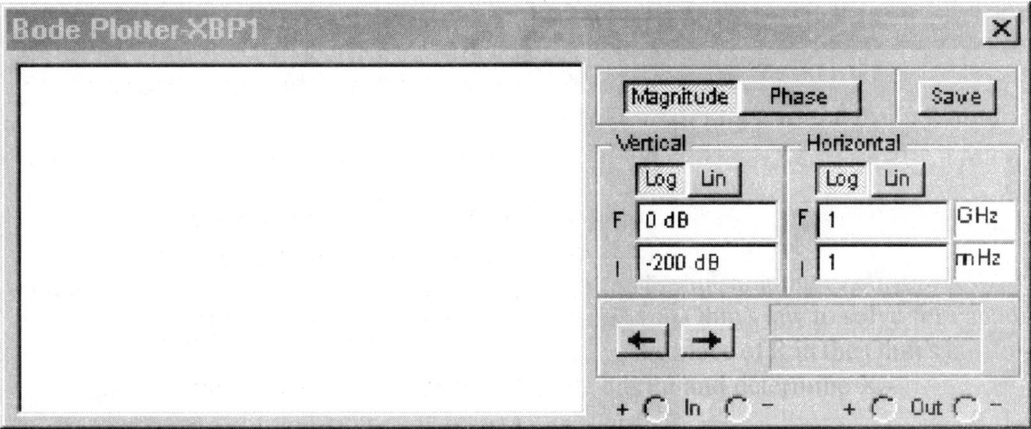


**Figure 18-1** The EWB Bode Plotter

2. At the top of the control panel to the right of the instrument, the **Magnitude** and **Phase** switches determine the type of measurement to be undertaken. The magnitude setting measures the ratio of magnitudes (voltage gain) between two points (V+ and V–). The phase setting measures the phase shift (in degrees) between two points. Both gain and phase shift are plotted against the signal frequency.

3. On the **Horizontal Axis** (1.0 mHz–10.0 GHz) the display shows the frequency range (in this case, one millihertz to 1 Gigahertz) to be measured. The I and F settings for the horizontal axis determine the scale of the frequency range. These ranges can be changed by clicking on the white space to the left of the GHz or mHz. The Log/Lin settings determine the frequency scale. A logarithmic scale is usually used in frequency response.

4. The units and scale for the **Vertical Axis** depend on what is being measured and the base being used as shown in Table 18-1.

| When Measuring | Using the Base | Minimum Initial Value | Maximum Final Value |
|---|---|---|---|
| Magnitude (gain) | Logarithmic | –200 dB | 200 dB |
| Magnitude (gain) | Linear | 0 | 10e + 09 |
| Phase | Logarithmic | –720° | 720° |
| Phase | Linear | –720° | 720° |

**Table 18-1** Range and Units for the Bode Plotter

5. The **Save** switch on the upper right section of the display panel allows you to save the Bode plot as a file on your hard drive or floppy disk.

6. The voltage and frequency for the cursor at any location on the display is shown in the two windows below the Horizontal section (bottom right).

7. Table 18-1 tells us that when measuring voltage gain, the vertical axis shows the ratio of the circuit's output voltage to its input voltage. For a logarithmic base, the measurement units are in decibels. For a linear base, the vertical axis shows the ratio of output voltage to input voltage. When measuring phase, the vertical axis always shows the phase angle between the input signal and the output signal in degrees. Regardless of the units, the initial (I) and final (F) values for the axis can be set using the Bode plotter controls.

8. Open circuit file **18-01a**. In this circuit, the Bode plotter is set to measure the voltage difference (magnitude) between the voltage applied to the circuit (TPA) and the voltage across the capacitor (TPB). Notice that the Vertical F and I settings are 0 dB and –20 dB respectively. The signal starts at 0 dB at the left of the screen and curves down to –20 dB to the right of the screen. Activate the circuit. Move the cursor to the right until you get it as close to the –3 dB point as possible.

9. Figure 18-2 shows the Bode plot for this circuit with the cursor set approximately at the –3db point (–2.992 dB). When it is necessary to observe a certain point on the plot, (concerning dB level and frequency, and so on), move the vertical cursor to the desired point on the Bode plot. You can also use the left-right arrows in the middle of the instrument to move the cursor. In this circuit, what is the frequency at the –3 dB point (–2.992 dB)?

   The frequency of this circuit at the –3 dB is ≈ __________ Hz (the ≈ symbol means approximately in mathematics). Move the cursor and observe the voltage and frequency variations. Notice that the voltage drop across the capacitor decreases as frequency increases.

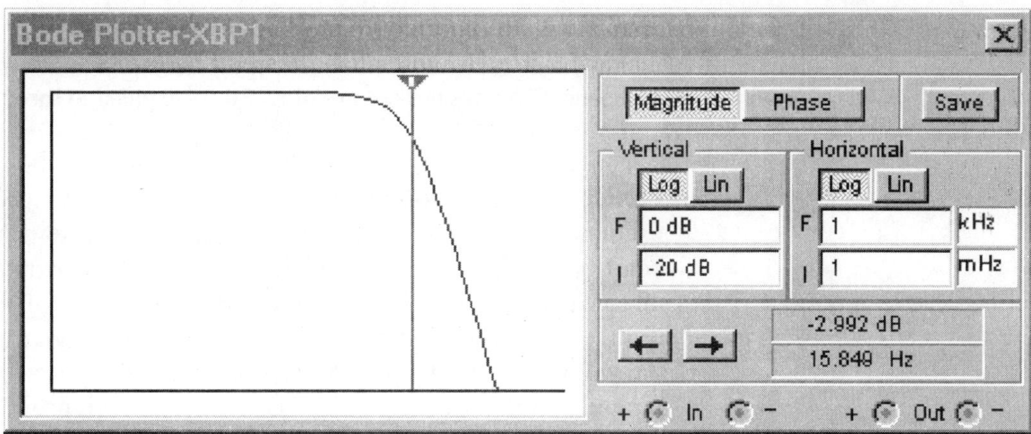


**Figure 18-2** A Bode Plot of an RC Circuit

10. Change the measurement mode from Magnitude to Phase. Move the cursor to the right until the –3 dB (–2.992 dB) frequency located in Step 9 is displayed in the cursor window at the bottom/right of the screen. What is the phase shift at –3 dB (–2.992 dB)?

    The phase shift is ___________ °.

11. Open circuit file **18-01b**. In this circuit an inductor has replaced the capacitor. It is expected that the voltage drop across the inductor will increase as frequency increases. Activate the circuit and observe that the frequency rises from some point out in the frequency spectrum. Move the cursor to the right and determine the frequency at the –3 dB point.

    The frequency of this circuit at –3 dB (–3.029) is ≈ __________ MHz. What is the phase shift at the previously measured frequency?

    The phase shift is __________ °.

12. Open circuit file **18-01c**. In this circuit the inductor and the capacitor are in parallel. Here, some interaction between the reactive components is to be expected. Activate the circuit and observe that the frequency rises to some center frequency and falls off past the center frequency point. What is the center (resonant) frequency of this circuit?

    The center frequency of this circuit at ≈ 0 dB is __________ kHz. What is the phase shift at the measured center frequency?

    The phase shift at the center frequency is __________ °.

## Activity 18.2: Low-Pass Filter Circuits

1. The **low-pass filter** circuit is a circuit that passes low frequencies and blocks high frequencies. The components and their placement in the

circuit determine the criteria of what is passed and what is blocked. The –3 dB point on a Bode plot is the point (by definition) at which higher frequencies do not pass or are blocked from passing through the filter circuit.

2. Open circuit file **18-02a**. This is the same type of RC circuit that was used in Step 8 of the previous activity except that the value of the components have been changed and it is now known that this type of circuit is a low-pass filter. Use the Bode plotter to determine the frequency at which the signal no longer passes through the circuit (this is the –3 dB point). This frequency is called the cutoff frequency. The formula to determine cutoff frequency (see Figure 18-3) for an RC low-pass filter is:

$$f_c = \frac{0.159}{RC}$$

**Figure 18-3** The Cutoff Frequency Formula for an RC Low-Pass or High-Pass Filter

Calculate the cutoff frequency for this circuit after measuring it with the Bode plotter.

The measured cutoff frequency for this circuit is __________ Hz.

The calculated cutoff frequency for this circuit is __________ Hz.
What is the phase shift at the cutoff frequency? The phase shift at the cutoff frequency is __________ °.

3. Open circuit file **18-02b**. This is an RL low-pass filter circuit that does the same job as the RC low-pass filter. Again, use the Bode plotter to determine the frequency at which the signal no longer passes through the circuit (this is the –3 dB point). The calculated cutoff frequency for this circuit is __________ Hz. The measured cutoff frequency for this circuit is __________ Hz. The formula to determine cutoff frequency (see Figure 18-4) for an RL low-pass filter is. Measure the phase shift at the cutoff frequency. The phase shift at the cutoff frequency is __________ °.

$$f_c = \frac{0.159\,R}{L}$$

**Figure 18-4** The Cutoff Frequency Formula for an RL Low-Pass or High-Pass Filter

- ***Troubleshooting Problem:***

4. Open circuit file **18-02c**. This is a low-pass filter circuit and should have a cutoff frequency of 6.485 Hz, but the output measured by the Bode plotter at the cutoff frequency is about –37 dB. What is wrong with the circuit?

   The problem with the circuit is that ______________________________

   ______________________________________________________.

## Activity 18.3: High-Pass Filter Circuits

1. A **high-pass filter** circuit blocks low frequency signals and passes high frequency signals. As in every filter circuit, the components and their position in the circuit determine the criteria of what is passed and what is blocked. The –3 dB point on this type of Bode plot is the point at which higher frequencies begin to pass through the filter.

2. Open circuit file **18-03a**. This circuit is an RL high-pass filter that is designed to block low frequencies and pass high frequencies. To calculate cutoff frequency for this type of circuit use the formula shown in Figure 18-4.

   The calculated cutoff frequency in this circuit is __________ kHz. Activate the circuit and measure the cutoff frequency with the Bode plotter. The

   measured cutoff frequency is __________ kHz. What resistance change could be made to this circuit to lower the cutoff frequency? Increasing / Decreasing (circle one) the resistance value of the resistor would lower cutoff frequency. Change the value of $R_1$ to the values indicated in Table 18-2, measure the approximate cutoff frequency for each value of $R_1$, and enter the data in Table 18-2.

| **Change $R_1$ Value to:** | 2 kΩ | 5 kΩ | 7.5 kΩ | 10 kΩ | 12 kΩ |
|---|---|---|---|---|---|
| Measured Cutoff Frequency | | | 12 kHz | | |

**Table 18-2** High-Pass Filter Circuit Cutoff Frequencies

3. Open circuit file **18-03b**. This circuit is an RC high-pass filter. To calculate cutoff frequencies for this type of circuit, use the formula shown in Figure 18-5.

The calculated cutoff frequency is __________ Hz. Activate the circuit and measure the cutoff frequency with the Bode plotter.

The measured cutoff frequency is __________ Hz.

$$f_c = \frac{0.159\,R}{L}$$

**Figure 18-5** The Cutoff Frequency Formula for an RL Low-Pass or High-Pass Filter

### ● *Troubleshooting Problem:*

4. Open circuit file **18-03c**. This circuit is an RC high-pass filter with a designed cutoff frequency of 265 Hz. There is a problem with the circuit. Activate the circuit and measure the cutoff frequency. The measured cutoff frequency is __________ kHz. Locate the problem. The problem is __________________________________________.

## Activity 18.4: Band-Pass Filter Circuits

1. **Band-pass filter** circuits block low frequency signals up to a certain cutoff point in the frequency spectrum and block high frequency signals that are beyond a higher cutoff frequency. A band of frequencies pass through the circuit and frequencies on either side of the band-pass frequencies are blocked. As in every filter circuit, the components and their position in the circuit determines the characteristics of the circuit. With band-pass filter circuits we are concerned about **bandwidth**, **selectivity**, and the actual **band-pass**. The sharper the response curve of a resonant circuit, the more selective it is. The less selective a band-pass circuit is, the wider the band of frequencies that is allowed to pass through the circuit, and the wider its resultant bandwidth. Bandwidth is the width of the frequency spectrum between the band-pass frequencies at the cutoff points on the frequency plot.

2. Open circuit file **18-04a**. This circuit is a series resonant band-pass filter. At resonance, the inductive and capacitive reactances are equal. Activate the circuit and determine the center frequency and the –3 dB band-pass points with the Bode plotter (which is approximate). Using the Bode plotter, the center frequency (resonant point) is located at __________ kHz. The approximate –3 dB points are at __________ kHz and __________ kHz for a bandwidth of approximately __________ kHz.

3. Open circuit file **18-04b**. This circuit is the same as the one we just worked on except the value of the load resistor $R_l$ has been changed to 100 Ω. Activate the circuit and notice the changed frequency plot on the Bode plotter. It has a narrow bandwidth compared to the previous circuit. Determine the approximate band-pass and the bandwidth with the Bode plotter.

   The band-pass is from __________ kHz to __________ kHz.

   The approximate bandwidth is __________ kHz.

4. Open circuit file **18-04c**. This circuit is a parallel resonant band-pass circuit that operates in a similar manner to the series resonant circuit. Activate the circuit and use the Bode plotter to determine the band-pass points, the center frequency, and the bandwidth. Using the Bode plotter,

   the band-pass points are at __________ Hz and __________ Hz, the bandwidth is __________ Hz, and the center frequency is __________ Hz.

- ***Troubleshooting Problem:***

5. Open circuit file **18-04d**. This circuit is a series resonant band-pass circuit that is not operating properly. The center frequency is about 1.2 kHz and should be about 2 kHz. The output is unable to get close to 0 dB; it is about –6.5 dB.

   The problem is ______________________________________________.

## Activity 18.5: Band-Reject, Bandstop, and Notch Filter Circuits

1. Bandstop, band-reject and notch filters prevent certain frequencies from passing through a circuit and are all titles for the same circuit action. Essentially, the methods are the same as for the band-pass filters, but the components are arranged differently to achieve the desired effect of blocking a band of frequencies rather than passing them.

2. Open circuit file **18-05a**. This is a series notch filter. Activate the circuit and measure the band-reject points, the band-reject width, and the center frequency of the notch with the Bode plotter. Using the Bode plotter, the

   band-reject points are at __________ Hz and __________ Hz, the band-reject width is __________ Hz, and the center frequency of the notch is __________ Hz.

3. Open circuit file **18-05b**. This is a parallel notch filter. Activate the circuit and measure the band-reject points, the band-reject width, and the center frequency of the notch with the Bode plotter.

   Using the Bode plotter, the band-reject points are at __________ Hz and __________ Hz, the band-reject width is __________ Hz, and the center frequency of the notch is __________ Hz.

## ● *Troubleshooting Problem:*

4. Open circuit file **18-05c**. This is the same circuit as Step 3. There is a problem with this circuit; the notch is poorly defined.

   The problem is ____________________________________________.